Contents

Contributors   4

Foreword   5

Preface   5

Overview of conclusions   6
Martin Parry, Nigel Arnell, Pam Berry, David Dodman, Samuel Fankhauser, Chris Hope, Sari Kovats, Robert Nicholls, David Satterthwaite, Richard Tiffin, Tim Wheeler, Jason Lowe, Clair Hanson

1. The range of global estimates   19
Samuel Fankhauser

2. Costs of adaptation in agriculture, forestry and fisheries   29
Tim Wheeler and Richard Tiffin

3. Costs of adaptation in the water sector   40
Nigel Arnell

4. Adaptation costs for human health   51
Sari Kovats

5. Adaptation costs for coasts and low-lying settlements   61
Robert Nicholls

6. The costs of adapting infrastructure to climate change   73
David Satterthwaite and David Dodman

7. Costing adaptation for natural ecosystems   90
Pam Berry

8. The costs and benefits of adaptation   100
Chris Hope

Contributors

Nigel Arnell, Walker Institute, University of Reading, UK
n.w.arnell@reading.ac.uk

Pam Berry, Environmental Change Institute, University of Oxford, UK
pam.berry@eci.ox.ac.uk

David Dodman, International Institute for Environment and Development, UK
david.dodman@iied.org

Samuel Fankhauser, Grantham Research Institute and Centre for Climate Change Economics and Policy, London School of Economics and Political Science, UK
s.fankhauser@lse.ac.uk

Clair Hanson, School of Development Studies, University of East Anglia, UK
clair.hanson@uea.ac.uk

Chris Hope, Judge Business School, University of Cambridge, UK
chris.hope@jbs.cam.ac.uk

Sari Kovats, London School of Hygiene & Tropical Medicine, UK
sari.kovats@lshtm.ac.uk

Jason Lowe, Meteorological Office, UK
jason.lowe@metoffice.gov.uk

Robert Nicholls, School of Civil Engineering and the Environment and the Tyndall Centre for Climate Change Research, University of Southampton, UK
R.J.Nicholls@soton.ac.uk

Martin Parry, Grantham Institute for Climate Change and Centre for Environmental Policy, Imperial College London, UK
martin@mlparry.com

David Satterthwaite, International Institute for Environment and Development, UK
david.satterthwaite@iied.org

Richard Tiffin, Walker Institute and Department of Agricultural and Food Economics, School of Agriculture, Policy and Development, University of Reading, UK
j.r.tiffin@reading.ac.uk

Tim Wheeler, Walker Institute and Department of Agricultural and Food Economics, School of Agriculture, Policy and Development, University of Reading, UK
t.r.wheeler@reading.ac.uk

Foreword

It is a pleasure to publish this assessment on behalf of a team of UK experts on adaptation to climate change. Its timing is crucial because, as the UNFCCC moves towards an agreement on post-Kyoto actions to meet the challenge of climate change, the role of funding for adaptation in middle- and low-income countries is taking a central place in negotiations. It therefore is imperative that our estimates of funding need are as accurate as possible, and we thank Professor Parry and his colleagues for taking us an important step further towards a robust costing of adaptation.

Camilla Toulmin
Director, International Institute for Environment and Development, London

Sir Brian Hoskins
Director, Grantham Institute for Climate Change, Imperial College, London

Preface

This is a preliminary evaluation based upon our own studies and without funding from any government or agency. Its conclusions are ours collectively and do not represent those of an organisation. We are grateful for comments, on an earlier draft, from authors of the original UNFCCC study.

Overview of conclusions

Martin Parry, Nigel Arnell, Pam Berry, David Dodman,
Samuel Fankhauser, Chris Hope, Sari Kovats,
Robert Nicholls, David Satterthwaite,
Richard Tiffin, Tim Wheeler

with contributions from Jason Lowe and Clair Hanson

Overview of conclusions

Summary

Several recent studies have reported adaptation costs for climate change, including for developing countries. They have similar-sized estimates and have been influential in discussions on this issue. However, the studies have a number of deficiencies which need to be transparent and addressed more systematically in the future. A re-assessment of the UNFCCC estimates for 2030 suggests that they are likely to be substantial under-estimates. The purpose of this report is to illustrate the uncertainties in these estimates rather than to develop new cost estimates, which is a much larger task than can be accomplished here. The main reasons for under-estimation are that: (i) some sectors have not been included in an assessment of cost (e.g. ecosystems, energy, manufacturing, retailing, and tourism); (ii) some of those sectors which have been included have been only partially covered; and (iii) the additional costs of adaptation have sometimes been calculated as 'climate mark-ups' against low levels of assumed investment. In some parts of the world low levels of investment have led to a current adaptation deficit, and this deficit will need to be made good by full funding of development, without which the funding for adaptation will be insufficient. Residual damages also need to be evaluated and reported because not all damages can be avoided due to technical and economic constraints. There is an urgent need for more detailed assessments of these costs, including case studies of costs of adaptation in specific places and sectors.

1 Introduction

This is an evaluation of estimates of the costs of adaptation made by the United Nations Framework Convention on Climate Change (UNFCCC) in 2007 and by some preceding studies (UNFCCC, 2007; Stern, 2006; World Bank, 2006; Oxfam, 2007; UNDP, 2007). The costs have been used as the basis for discussion regarding the levels of investment needed for adaptation to climate change. They have been influential in the debate concerning funding for climate change and it is important, therefore, that such estimates of cost are as robust as possible. The purpose of this report is to assess these estimates and consider ways to improve them in the future.

The UNFCCC report was based on a set of commissioned studies (UNFCCC background papers, 2007). These took place over a short period dictated by the timescale of the UNFCCC process and the need to report the results to the next Conference of the Parties, so there was no time for independent review of a draft of the report.

It is important, therefore, to recall the objectives of the UNFCCC report and the caveats that the authors ascribed to its conclusions. The study was a preliminary one of the funding, especially the public funding, estimated to be needed in the year 2030 to meet the challenge of climate change. It is not a study of the full cost of avoiding all damage. It does not cover some important activities, and other activities are only partially covered. The authors suggest that their estimates are probably under-estimates and that much more study is needed.

The purpose of this evaluation is to consider the relative strengths and weaknesses of the UNFCCC study, so that we can determine what next steps can be taken to improve our understanding of the issue. It is not our purpose here to develop a revised set of numbers for the funding of adaptation to climate change, because we believe this requires detailed study.

Overview of conclusions

But we do conclude that the UNFCCC study has probably under-estimated this cost, and we take this under-estimate to be substantial. The next step needs a substantial study combining 'bottom-up' local case studies with 'top-down' integration, which is more than is possible in this report. We conclude with some recommendations for that study.

2 Studies before the UNFCCC study

The UNFCCC estimates of adaptation cost are broadly in line with preceding studies published by the World Bank, Oxfam, UNDP, and in the Stern report. These have recently been summarised by the OECD and are given in Table 1. Since these studies appear to support each other, the conclusion has sometimes been made that there exists a comforting convergence of evidence, but that would be misleading because: (i) none of these are substantive studies, (ii) they are not independent studies but borrow heavily from each other, and (iii) they have not been tested by peer review in the scientific or economics literature.

Table 1 Estimates of adaptation costs in developing countries, for 2010-2015

Source	US$ billion p.a.	Comments
World Bank (2006)	9-41	Cost of climate-proofing FDI, GDI and ODA flows
Stern (2006)	4-37	Update, with slight modification of World Bank (2006)
Oxfam (2007)	>50	Based on World Bank, plus extrapolation of costs from NAPAs and NGO projects
UNDP (2007)	86-109	World Bank, plus costing of PRS targets, better disaster response

Source: Agrawala and Fankhauser (2008)
Note: FDI = foreign direct investment, GDI = gross domestic investment, ODA = official development assistance, NAPA = National Adaptation Programme of Action, PRS = poverty reduction strategy

Importantly, most of the precursors of the UNFCCC study were based on the same method, first developed by the World Bank (2006). This takes a fraction of current investment that is climate-sensitive and applies a 'mark-up' factor to this fraction to reflect the cost of 'climate-proofing' those investments. We shall consider below the weakness of this approach.

3 The UNFCCC study

The UNFCCC commissioned six studies which provided estimates of the cost of adaptation for the year 2030, usually assuming a climate scenario similar to the IPCC's SRES A1B and B1. In summary these cover:

- **Agriculture, forestry and fisheries.** The agriculture estimate (McCarl, 2007) consists of three distinct cost items: extra capital investment at farm level, the need for better extension services at country level and the cost of additional global research (e.g. on new cultivars).

Overview of conclusions

- **Water supply.** The water estimate (Kirshen, 2007) considers the effect of additional water demand and changes on the supply side. Investment decisions are made in anticipation of 2050 water needs.
- **Human health.** The health estimates (Ebi, 2007) are the extra prevention costs for three health issues: malnutrition, malaria and diarrhoea. The health impacts are based on the Global Burden of Disease study (McMichael et al., 2004).
- **Coastal zones.** Coastal protection costs are based on the DIVA model (Nicholls, 2007), which considers a limited set of adaptation options that are applied globally. Uniquely, the coastal estimate considers both adaptation costs and residual damages. For long-life defence infrastructure, investments are made in anticipation of sea-level rise in 2080.
- **Infrastructure.** The infrastructure estimate adopts the World Bank (2006) methodology, using insurance data to determine the share of climate-sensitive investment, and applying a percentage increase on current infrastructural investment to suggest additional costs for climate-proofing new infrastructure. (However, the background paper by Satterthwaite (2007) took a different approach.)
- **Ecosystems.** An indication of adaptation costs for ecosystems was derived from the costs of increasing protected areas to at least 10% of the land area of each nation or ecosystem, although it was not possible to split this into baseline costs of meeting current deficits and incremental adaptation. See Berry (2007).

The UNFCCC report concluded that total funding need for adaptation by 2030 could amount to $49 – 171 billion per annum globally, of which $27 – 66 billion would accrue in developing countries (Table 2) (note that all references to dollars in this report are to US dollars unless otherwise specified). By far the largest cost item is infrastructure investment, which for the upper-bound estimate accounts for three-quarters of total costs. Costs are over and above what would have to be invested in the baseline to renew the capital stock and accommodate income and population growth. Note that the total excludes the estimate for ecosystem adaptation, for reasons discussed below in this report.

Table 2 UNFCCC estimate of additional annual investment need and financial flow needed by 2030 to cover costs of adaptation to climate change (billion dollars per year in present-day values)

Sector	Global cost	Developed countries	Developing countries
Agriculture	14	7	7
Water	11	2	9
Human health	5	Not estimated	5
Coastal zones	11	7	4
Infrastructure	8 – 130	6 – 88	2 – 41
Total	**49 – 171**	**22 – 105**	**27 – 66**

Source: UNFCCC (2007)

Overview of conclusions

4 Issues affecting the robustness of the estimates

In reviewing the UNFCCC report (and others), the following issues emerge which require more thorough treatment in future assessments.

The potential damages to be avoided by adaptation

The Fourth Assessment Report of the IPCC (IPCC, 2007) gives a summary of some impacts likely to occur under varying amounts of global warming. Mapping onto this is the expected warming range for 2030 which indicates the potential impacts that adaptation will need to address. Figure 1 shows this for the A1B scenario assumed generally in the UNFCCC study, as well as for 2050 and 2080 (used respectively in the water and coastal analyses of that study, on the assumption that adaptation in 2030 will need to anticipate future warming due to the long-term nature of investment needs in those sectors). One aspect emerging here is the substantial magnitude of impacts that could occur even within the next few decades, and the scale of damages that could be expected if adaptation is not fully successful in avoiding them.

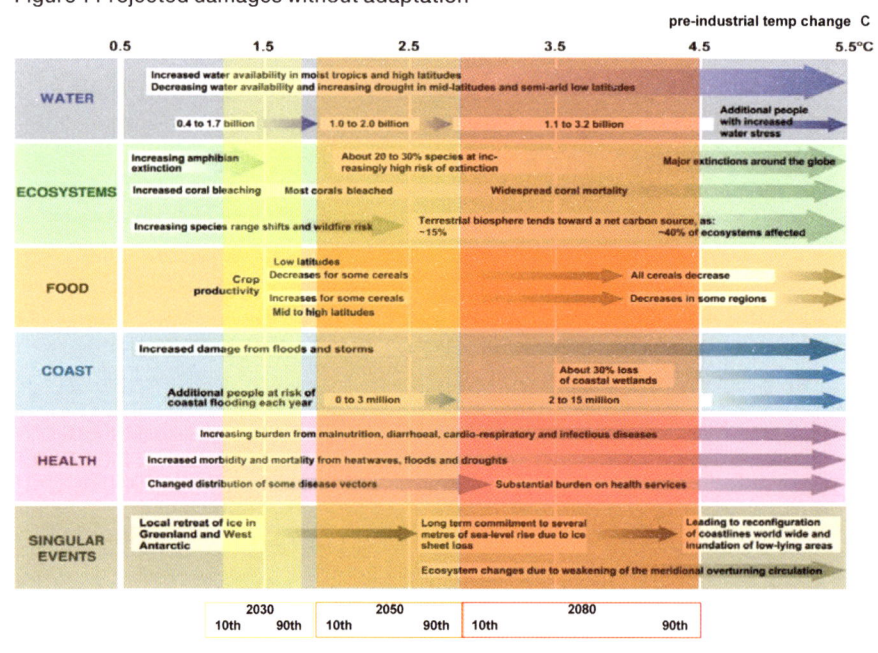

Figure 1 Projected damages without adaptation

Source: This background table of impacts is from the Technical Summary of IPCC Working Group II (IPCC, 2007)

Note: The shaded columns show the 10 and 90 percentile uncertainty range for the scenarios assumed in the UNFCCC estimation of funding needs for adaptation. For most sectors the A1B scenario was taken. For water and coastal protection the scenarios were 2050 and 2080 respectively, due to the need for adaption to anticipate future climate change. Scenario data from a simple Earth system model (Parry, Lowe and Hanson, 2009)

Overview of conclusions

The scarcity of information on adaptation and its cost

Information is scarce about the scale of future potential impacts, and is even more scant for the costs of avoiding them by adaptation, a point stressed in the UNFCCC report. Some sectors such as mining and manufacturing, energy, retailing, and tourism, were not included in the UNFCCC report. Cost estimations for ecosystems, although made, were left out of the final table showing total costs (see Table 2 above), due to lack of sufficiently robust information. Within some sectors that were examined, the funding needs estimated were clearly only partial. In health for example, just three areas of impact, where there were sufficient estimates, were considered: the effect of climate change on diarrhoeal diseases, malaria and malnutrition in low- and middle-income countries. Adaptation costs for health effects in high-income countries were not estimated.

A major problem is the absence of case studies to test the top-down form of UNFCCC analysis. The few national figures available tend to suggest costs in excess of the UNFCCC estimates. For example, agencies responsible for flood management in England and Wales have estimated a need to spend (due to climate change) an additional $30 million annually in 2011, growing to $720 million by 2035 (Environment Agency, 2009).

Applying a 'climate mark-up' against future investment trends

In most cases, the UNFCCC estimation of funding needs was derived from applying an increase in cost to those areas of investment that are deemed to be climate-sensitive. In agriculture for example, 2% of investment on infrastructure is taken to be climate-sensitive. In some sectors, particularly the built environment, the investment flows are so large that even small changes in this mark-up can change estimates significantly.

Investment needed to remove the 'adaptation deficit'

In particular, applying a 'climate mark-up' is not appropriate when current investment flows are well below what they should be. In several parts of the world, current levels of investment are considered far from adequate, and lead to high current vulnerability to climate, including its variability and extremes, the latter case being termed a current 'adaptation deficit' (Burton, 2004). This partly explains why impacts from climate change are expected to be greatest in low- and middle- income countries (IPCC, 2007). To avoid these impacts the adaptation deficit (which is largely a development deficit) will need to be made good. For good reason these costs were not included in the UNFCCC estimate, which was aimed at identifying the additional cost of climate change, but it needs to be stressed that without the adaptation deficit being made good, the enhanced investment for adaptation will not be sufficient to avoid serious damage from climate impacts. Dlugolecki (2007), in a background paper for the UNFCCC study, estimated (at $200 billion per year) the costs of damage from present-day extreme weather and took this as a reflection of the current scale of inadequate adaptation. The Millennium Development Goals represent an attempt to make good some, but probably not all, of the adaptation deficit, and have been costed at about $200 billion by 2015 (Sachs and McArthur, 2005). To make good the full development deficit probably requires enhancing official development assistance to 0.7% of GDP of OECD countries. Hence the issues of development and adaptation costs are intimately linked, and this requires further exploration.

Overview of conclusions

Adaptation costs in a world without an 'adaptation deficit'

There follows from this the question as to whether the 'climate mark-up' should be against investment levels that reflect current trends (which in many regions, of Africa for example, are insufficient to remove high levels of vulnerability to climate), or to elevated levels that are needed to attain the MDGs, or to even higher levels that would help achieve sustainable and equitable development. The UNFCCC takes the former line, but this leads to estimations of cost that are substantially lower than if one assumed a development pathway which protects the poor against vulnerability to climate change. In this report we conclude that removing the housing and infrastructure deficit in low- and middle-income countries will cost around $315 billion per year (in today's figures) over 20 years; while adapting this upgraded infrastructure specifically to meet the challenge of climate change will cost an additional $16–63 billion per year.

How much impact is being avoided by adaptation?

It is not clear what proportion of expected damage would be avoided by the proposed UNFCCC investment levels. Most impacts are projected to increase non-linearly with climate change, and adaptation costs similarly with impacts (IPCC, 2007). Therefore it will probably be very inexpensive to avoid some impacts but prohibitively expensive to avoid others; and some impacts we cannot avoid even if funds were unlimited, because the technologies are not available (e.g. in connection with ocean acidification). A simple schema of a generalised adaptation cost curve is shown in Figure 2. The curve is likely to vary greatly between different sectors and places, but probably common to most cases will be that adaptation to (say) the first 10% of damage will be disproportionately cheaper than for 90% of damage. We need to be clear, then, about how much we are willing to pay for adaptation to avoid damages. To illustrate, we might aim (in a scale of reducing cost) to adapt to: (i) all those impacts that reduce human welfare, or (ii) all those that are economically feasible (i.e. cheaper to adapt to than to be borne), or (iii) all those that are affordable within a given budget constraint (for example, the size of the global Adaptation Fund).

Figure 2 Schematic of adaptation costs, avoided damages and residual damage compared: a) at a point in time, and b) over time

Figure 2a

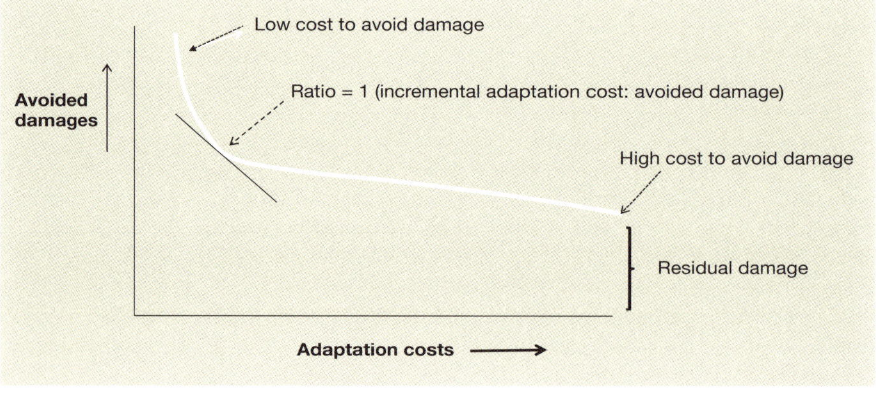

Overview of conclusions

Figure 2b

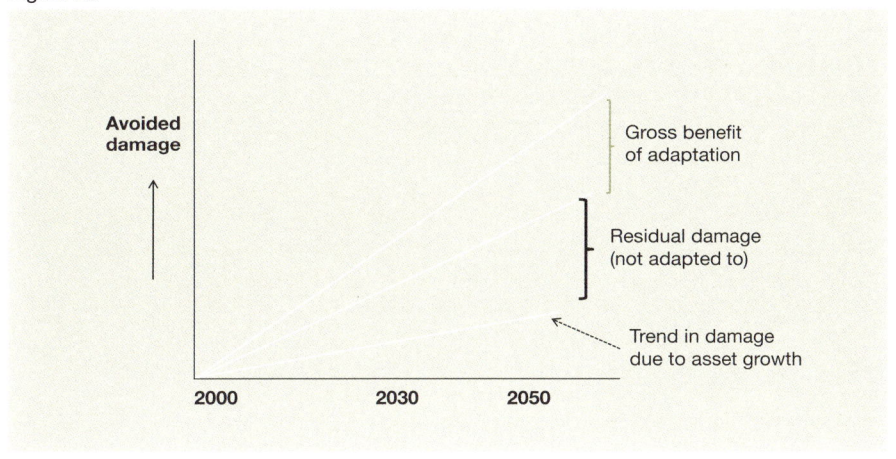

[Figure 2b: Graph showing Avoided damage on y-axis from 2000 to 2050 on x-axis, illustrating Gross benefit of adaptation, Residual damage (not adapted to), and Trend in damage due to asset growth]

The costs of damage not adapted to, or 'residual damage'

Implicit in the above (and illustrated in Figure 2) is that much damage will not be adapted to over the longer term, because adaptation is either not economic or not feasible. We term this 'residual damage'. In the UNFCCC report it is not clear how much residual damage might be expected. But it is very important that we start to consider this, because the amount may be significant and is likely to increase over time. In the evaluation reported here, residual impacts are estimated at about a fifth of all impacts in agriculture in 2030 (see Chapter 2 in this report) and, over the longer term, may account for up to two-thirds of all potential impacts across all sectors, depending on the amount of climate change not avoided by mitigation (see Chapter 8 in this report).

'Soft adaptation'

The UNFCC study may have given insufficient weight to the value of 'soft adaptation'. It is easier analytically to cost out structural measures like the expansion of water supply systems, and the UNFCCC study focused on these. In reality, it will often be cheaper to apply 'soft adaptation' options. Measures to use water more efficiently, for example, may obviate the need for expensive new infrastructure. Conversely, the human health costs do not include changes in infrastructure ('hard adaptation') which may be considerable (see Chapter 4 in this report).

How will adaptation costs change over time?

The UNFCCC estimate of adaptation costs is a 'snapshot' for 2030 at one point along the climate-impact curve, and its authors note the importance of the question, 'While the adaptation cost curve seems quite gentle between now and 2030, how steeply will it grow thereafter?' Some believe it may rise steeply, possibly quadratically in some sectors (IPCC, 2007). It is very important that this be analysed so we are sufficiently prepared for escalating adaptation costs beyond 2030.

Overview of conclusions

The level of under-estimation in the UNFCCC study

For a number of reasons discussed above, and given in more detail in the following section, we believe that the UNFCCC estimate of investment needs is probably an under-estimate by a factor of between 2 and 3 for the included sectors. It could be much more if other sectors are considered. We conclude that for coastal protection the factor of under-estimation could be 2 to 3. For infrastructure it may be several times higher, at the lower end of the cost range. For health the 'intervention sets' that were costed relate to a disease burden that is approximately 30–50% of the anticipated total burden in low- and middle-income countries (and do not include interventions in high-income countries). Including ecosystems protection could add a further $65–$300 billion per year in costs. Furthermore, estimates are not made for sectors such as mining and manufacturing, energy, the retail and financial sectors and tourism. This probably explains why the investment levels proposed by the UNFCCC appear so small – roughly the annual cost of running two or three Olympic Games. That this represents a doubling of current ODA only highlights the very low current level of development assistance.

Recommendations for future studies

It is important that robust studies of adaptation cost are, in future, based upon case studies that cover a wide range of places and sectors, and support top-down analyses of the kind evaluated here. The World Bank and McKinsey will be shortly reporting on this (World Bank, 2009; McKinsey, 2009). The time period and expected climate changes need specifying (as they were in the UNFCCC study), and results for multiple timeframes would be useful. Non-climate trends need careful portrayal, especially the future levels of non-climate investment. Costs of adapting to varying amounts of impact should be analysed, thus providing a choice range for preparedness to pay; and there needs to be some analysis of the residual impact that adaptation is not likely to avoid, and the resulting damage costs that we need to anticipate.

5 Summary of conclusions for each sector

This section presents summary conclusions from Chapters 2–7 below, considering costs of adapting to climate changes in the six areas of agriculture, water, human health, coasts, infrastructure and natural ecosystems.

Agriculture

The UNFCCC estimated the cost of adaptation of agriculture to climate change at $11.3–12.6 billion for 2030. The basis for this is an assumption that the 'climate mark-up' will amount to 10% of research and extension spend, and 2% of infrastructure spend. In the background paper the authors state that these assumptions are uncertain and speculative given the limited basis from which they were formed, but the UNFCCC report does not repeat these caveats nor make clear the reasoning behind the mark-up levels that were adopted.

Overview of conclusions

One outcome of the small 'climate mark-up' is the UNFCC conclusion that the cost of adapting to climate change will be one-fortieth the cost of adapting to population change (i.e. meeting increased demand). This contrasts with estimations for the water sector where the ratio is given as 1:3. These are important differences which deserve analysis.

The current adaptation deficit in agriculture is high. The number of people at risk from hunger increased from around 300 million in 1990 to 700 million in 2007, and may exceed 1 billion in 2010 (FAO, 2009). A measure of the cost of making good this deficit is the cost of achieving the relevant Millennium Development Goal, estimated at $40–60 billion per year. Without this non-climate investment, the estimated levels of investment for adaptation to climate change will be insufficient to avoid serious damage.

There are a few bottom-up case studies which indicate the magnitude of adaptation costs, and these suggest that UNFCCC may be on the low side of adaptation costs in this sector. For example, there is an estimate of $8 billion for adapting crop irrigation systems to climate change by 2030, and of $14.5 billion for the year 2030 for the reduction in the value of global crop outputs due to climate change.

For these reasons the present study concludes that the UNFCCC estimate of $11.3–12.6 billion is a reasonable first approximation of adaptation costs in this sector, but we expect the estimate of adaptation costs for agriculture, forestry and fisheries to increase as more detailed studies of specific adaptation actions become available. Finally, the present study concludes that, with such levels of adaptation, about 80% of the cost of potential impacts might be avoided, but about 20% might not.

Water

The UNFCCC estimated national water resource availability in relation to large-area projections of national rainfall, and then compared availability with expected demand. It assumed that a quarter of the total cost of adaptation due to changes in demand and supply would be due to changes in supply resulting from climate change (i.e. $11 billion per year). This 1:3 ratio contrasts with 1:40 given by the UNFCCC for agriculture.

The study worked at the national level only, assuming that water resources could be transferred within a country from areas of surplus to areas of deficit. For small countries this would not be a problem, but for large ones it is unrealistic and probably a source of under-estimation of true cost. To illustrate, for a single basin in China (Huang Ho) the annual costs of adapting to climate change could be $0.5 billion per year (see Chapter 3 in this report). Unfortunately, few such studies are currently available and are insufficient for drawing reliable conclusions.

The UNFCCC costs include that of water provision, but not of adapting to altered flood risk in river basins. These altered flood management costs may be very substantial (potentially $0.1–0.2 billion annually in the Sacramento Basin in California alone), but there have been no consistent inter-comparisons of costs in different parts of the water sector. The use of an average climate change scenario rather than an ensemble which describes the range of possible impacts has probably led the UNFCCC to under-estimate the costs of providing for the full range water-storage need. In all, these costs omitted from the UNFCCC could be very substantial.

Overview of conclusions

Human health

The UNFCCC estimates of costs of adaptation are the costs of the intervention set to prevent the additional burden of disease due to climate change for three health outcomes in low- and middle-income countries: diarrhoeal diseases, malaria and malnutrition. The estimates are in the range of $4–12 billion per year in 2030.

These three outcomes are not the total projected burden on human health from climate change. That total has yet to be assessed accurately, but authors of the WHO study of the global disease burden estimate that these outcomes amount to 30–50% of the probable future total burden in 2030 in low- and middle-income countries (McMichael and Bertollini, 2009, personal communication).

A potential source of under-estimation is that the UNFCCC considers a narrow range of development futures. It takes a single median population projection in which population numbers increase and cases of diarrhoea/malaria/malnutrition remain constant, i.e. there is steep relative decline in incidence. The present study considers this to be an optimistic assumption.

Coasts

The costs reported by the UNFCCC study for coastal adaptation appear reasonable as a snapshot cost for IPCC sea-level rise projections in 2030, and are more reliable than those for other sectors since they are based on a model assessment rather than top-down assumptions. However, some post-2007 IPCC assessments suggest significantly higher observed and projected rises in sea levels than reported in the 2007 IPCC assessment. If these are assumed, then adaptation costs would be roughly doubled.

If account is taken of the need to protect coastal landscapes for amenity or ecological reasons, then the adaptation approach might change and protection costs could increase significantly in most cases. Another deficiency of the UNFCCC study is the lack of consideration of other aspects of climate change such as more intense storms. No detailed estimates are available, but in the worst-case the necessary adaptation costs (and residual damage costs) could match those of adapting to sea-level rise.

When combined with the uncertainties about sea-level rise, adaptation costs three times those reported in the UNFCCC study are not implausible. Experience with the DIVA model since Nicholls (2007) suggests that UNFCCC estimations for residual damages are overly conservative and should be roughly doubled (to $2–$3billion per year).

Infrastructure

To estimate adaptation costs for infrastructure, the UNFCCC: (i) estimated global investment in gross fixed capital formation in 2030 (around three times the global investment in 2000), as $22.2 trillion; (ii) multiplied this by the proportion that is vulnerable to the impact of climate change (based on data for losses from weather disasters – 0.7% (Munich Re data) or 2.7% (ABI data)), resulting in $153–650 billion a year; then (iii) assumed 5–20% of this total as the increase in capital costs needed for adaptation, giving $8–31 billion (Munich Re) or $33–130 billion (ABI data).

Overview of conclusions

The estimates based on the Munich Re data are likely to be substantial under-estimates of damages from climate because only data from large events are included and (as the UNFCCC study notes), this leaves out the cost of a high proportion of all extreme-weather disasters. The climate-cost fraction of 0.7–2.7% was recognised as being low by the authors of the background papers to the UNFCCC report, and in earlier work they adopted climate-cost fractions of 2–10% for domestic investment and up to 40% for overseas development assistance.

Applying a climate mark-up to levels of infrastructure provision that are currently very low (e.g. in most countries in Africa, many in Asia, and considerable parts of Latin America and the Caribbean) yields low estimations of future cost. Infrastructure provision needs substantial improvement to meet present-day needs and these are partly embraced in the Millennium Development Goals. The present study concludes that removing the housing and infrastructure deficit in low- and middle-income countries will cost around $315 billion per year (in today's figures) over 20 years. Adapting this upgraded infrastructure specifically to meet the challenge of climate change will cost an additional $16–63 billion per year.

Investment in adaptation will not avoid all damages to infrastructure. The annual economic damage caused by large extreme-weather disasters, 1996–2005, was over $50 billion a year. This gives an indication of weather impacts that are currently not avoided by adaptation – even in countries where the population is served by protective infrastructure and good-quality buildings.

Ecosystems

The UNFCCC methodology consisted of estimating: (i) the current global expenditure on conservation in the form of protected areas (PAs), (ii) the shortfall in the PA network (PAN), (iii) the level of additional expenditure needed for PAs to be adequate for climate-change adaptation, and (iv) costing adaptation outside the protected area network. An estimate of $12–22 billion was given by UNFCCC for the cost of expanding and protecting terrestrial PAN areas so that they represent 10% of each country, but this was excluded from the ultimate list of needed investment.

The present study supports the conclusion of the background paper for the UNFCCC, which is that $65–80 billion reflects the range of probable adaptation costs for PAN areas, including both terrestrial and marine environments. Additionally, this study supports the conclusion in the same background paper that adaptation costs for non-protected (non-PAN) areas should be included and could amount to about $290 billion, although these involve key assumptions and have a higher degree of uncertainty than estimates for some of the other fields. Since the UNFCCC report on global costs of adaptation omitted the costs of protecting ecosystems and the services they can provide for human society, the present study concludes that this is an important source of under-estimation.

Overview of conclusions

References

Agrawala, S. and Fankhauser, S. (eds) (2008). *Economic Aspects of Adaptation to Climate Change. Costs, Benefits and Policy Instruments*. OECD, Paris.

Berry, P. (2007). *Adaptation Options on Natural Ecosystems*. A report to the UNFCCC Financial and Technical Support Division, Bonn (http://unfccc.int/cooperation_and_support/financial_mechanism/ financial_mechanism_gef/items/4054.php).

Burton, I. (2004). 'Climate change and the adaptation deficit' in A. French, et al., *Climate Change: Building the Adaptive Capacity*, Meteorological Service of Canada, Environment Canada, pp25–33.

Dlugolecki, A. (2007). *The Cost of Extreme Events in 2030*. A report to the UNFCCC Financial and Technical Support Division (http://unfccc.int/cooperation_and_support/financial_mechanism/ financial_mechanism_gef/items/4054.php).

Ebi, K. (2007). *Health Impacts of Climate Change*. A report to the UNFCCC Financial and Technical Support Division (http://unfccc.int/cooperation_and_support/financial_mechanism/ financial_mechanism_gef/items/4054.php).

Environment Agency (2009). *Investing for the Future Flood and Coastal Risk Management in England: A Long-term Investment Strategy*. Environment Agency, Bristol (downloadable at http://publications.environment-agency.gov.uk/pdf/GEHO0609BQDF-E-E.pdf).

FAO (2009). '1.02 Billion People Hungry' (www.fao.org/news/story/en/item/20568/icode/) and (forthcoming, 2009) *The State of Food Insecurity in the World*.

IPCC (2007). *Climate Change 2007: Impacts, Adaptation and Vulnerability. Contribution of Working Group II to the Fourth Assessment Report of the Intergovernmental Panel on Climate Change*. M.L. Parry, O.F. Canziani, J.P. Palutikof, P.J. van der Linden and C.E. Hanson (eds), Cambridge University Press, Cambridge, 976pp.

Kirshen, P. (2007). *Adaptation Options and Costs in Water Supply*. A report to the UNFCCC Financial and Technical Support Division (http://unfccc.int/cooperation_and_support/ financial_mechanism/financial_mechanism_gef/items/4054.php).

McCarl, B. (2007). *Adaptation Options for Agriculture, Forestry and Fisheries*. A report to the UNFCCC Financial and Technical Support Division (http://unfccc.int/ cooperation_and_support/financial_mechanism/financial_mechanism_gef/items/4054.php).

McKinsey (2009) (forthcoming). *The Economics of Adaptation*.

McMichael, A.J., Campbell-Lendrum, D., Kovats, R.S., Edwards, S., Wilkinson, P., Edmonds, N., Nicholls, N., Hales, S., Tanser, F.C., Le Sueur, D., Schlesinger, M. and Andronova, N. (2004). 'Climate Change' in M. Ezzati et al. (eds), *Comparative Quantification of Health Risks: Global and Regional Burden of Disease due to Selected Major Risk Factors. Vol. 2*. World Health Organization, Geneva, pp1543–1649.

Nicholls, R. (2007). *Adaptation Options for Coastal Zones and Infrastructure*. A report to the UNFCCC Financial and Technical Support Division (http://unfccc.int/ cooperation_and_support/financial_mechanism/financial_mechanism_gef/items/4054.php).

Oxfam (2007). *Adapting to Climate Change. What is Needed in Poor Countries and Who Should Pay*? Oxfam Briefing Paper 104.

Parry, M.L., Lowe, J.A. and Hanson, C. (2009). 'Overshoot, Adapt and Recover', *Nature* 258(7242), pp1102–1103.

Sachs, J.D. and McArthur, J.W. (2005). 'The Millenium Project: A Plan for Meeting the Millennium Development Goals', *Lancet* 365, pp347–353.

Satterthwaite, D. (2007). *Adaptation Options for Infrastructure in Developing Countries*. A report to the UNFCCC Financial and Technical Support Division (http://unfccc.int /cooperation_and_support/financial_mechanism/financial_mechanism_gef/items/4054.php).

Stern, N. (2006). *Stern Review: Economics of Climate Change*. Cambridge University Press, Cambridge.

UNDP (2007). *Human Development Report 2007/08*. Palgrave McMillan, New York.

UNFCCC (2007). *Investment and Financial Flows to Address Climate Change*. Climate Change Secretariat, Bonn.

UNFCCC background papers (2007): see individual authored items cited in this list.

World Bank (2006). *Investment Framework for Clean Energy and Development*. World Bank, Washington DC.

World Bank (2009) (forthcoming). *The Economics of Adaptation to Climate Change*.

1

The range of global estimates

Samuel Fankhauser

1

Summary

- Global adaptation cost estimates range from $4 billion a year to well over $100 billion. These numbers refer to the annual cost of adapting to 'median' climate change over the next 20 years. They are different from estimates of the social costs of climate change, which measure adaptation costs plus residual impacts (that is, damages after adaptation) over the atmospheric lifetime of greenhouse gases.

- The wide range is symptomatic of the poor state of knowledge, with most estimates indicative and incomplete, but also of the analytical difficulty of defining adaptation. There is also a dearth of independent studies using different estimation techniques.

- The assumptions made to derive these estimates create both negative and positive biases, but on balance it is likely that current approaches under-estimate the true cost of adaptation.

1.1 Adaptation and the total costs of climate change

To understand the costs of adaptation one has to look at adaptation in its larger context. Adaptation is only one part of the overall response to (and therefore the costs of) climate change. The total burden of climate change consists of three elements: the costs of mitigation (reducing the extent of climate change), the costs of adaptation (reducing the impact of change), and the residual impacts that can be neither mitigated nor adapted to.

For example, society may seek to limit the overall temperature increase to 2°C (mitigation), invest in coastal protection to limit the negative impacts of 2°C warming (adaptation) and accept the loss of certain coastlines because they cannot be defended at reasonable cost (residual damage). In the simplest possible economic framework, society would fine-tune global mitigation and adaptation efforts until the combined (opportunity) costs of mitigation, adaptation and residual damage are minimised.[1]

What is important to understand at this point is that this optimisation process is more complex than simply comparing annual adaptation costs – as reviewed in this chapter – with estimates of global mitigation costs. There are two main complications. First, adaptation will not reduce impacts to zero. There may be substantial residual damages that have to be taken into account. Second, greenhouse gases are stock pollutants with long atmospheric lifetimes. Cutting emissions today reduces the need for costly adaptation (and residual impacts) not just today but over many decades – as long as the gases would have remained in the atmosphere.[2]

1.2 Adaptation and development

Another important delineation is between adaptation and socio-economic development. Socio-economic trends over the coming decades – population growth, economic expansion, the deployment of new technologies – will affect our vulnerability to climate events and indeed may be shaped by climate conditions. Human activity has always been influenced by the climate conditions people find themselves in. It is therefore difficult to delineate where socio-economic development ends and adaptation to anthropogenic climate change begins.

This is particularly the case for developing countries, where there is a well-documented adaptation deficit – that is, insufficient adaptation to the current climate. Poor people and poor countries are less well prepared to deal with current climate variability than rich people and rich countries. There is evidence that higher measures of development indicators like per capita income, literacy and institutional capacity are associated with lower vulnerability to climate events (Noy, 2009; Bowen et al., 2009). This has led authors like Schelling (1992) to conclude that good development is one of the best forms of adaptation. More subtly, McGray et al. (2007) identify a continuum of measures that address, to varying degrees, both development and adaptation needs. They range from measures that reduce vulnerability to stress more broadly (whether climate-related or not) to the creation of 'systems for problem solving', the management of current climate risks, and policies specifically addressing climate change.

Most cost estimates deliberately and understandably ignore the overlap between adaptation and development and focus on incremental adaptation over and above a vaguely defined baseline that includes climate-relevant development programmes. With these caveats in mind we turn to a review of existing adaptation cost estimates.

1.3 Adaptation costs in the impact literature

Research on adaptation costs has a surprisingly long history, starting with early attempts to estimate the economic costs of climate change.[3] The objective at that time was not to measure adaptation costs per se, but to refine our understanding of climate change impacts. Modellers recognised that their impact estimates would be wrong if they did not include an adaptive response and overcame the 'dumb farmer hypothesis' (the assumption that farmers and other actors would not react to a change in climate).

In a survey of adaptation in impact models, Tol et al. (1998) concluded that many impact categories covered in the economic cost literature were actually adaptation costs, in particular coastal protection, space heating and cooling (an adaptive response to changing temperatures), defensive expenditures against air pollution and in some cases migration (an adaptive response if premeditated but arguably a residual damage in the case of climate refugees). Adaptation also featured prominently in the agriculture literature and to a lesser extent in health, but the adaptive measures considered there were rarely costed out. Overall, Tol et al. found that adaptation costs amounted to 7–25% of total impacts, with residual damages accounting for the balance.

Over the years, the treatment of adaptation in global impact models was refined in a series of sector studies.[4] However, a recent survey found that beyond coastal protection our knowledge of adaptation costs (and benefits) is still fairly limited (Agrawala and Fankhauser, 2008), as shown in Table 1.1.

Table 1.1 The state of knowledge on adaptation costs and benefits

	Analytical coverage	Cost estimates	Benefit estimates
Coastal zones	Comprehensive	√√√	√√√
Agriculture	Comprehensive	–	√√
Water	Isolated case studies	√	√
Energy	N. America, Europe	√√	√√
Infrastructure	Cross-cutting, partly covered in other sectors	√√	–
Health	Selected impacts	√	–
Tourism	Winter tourism	√	–

Source: Agrawala and Fankhauser (2008)

1.4 Adaptation costs in country case studies

Other than in sector studies, adaptation costs are increasingly studied at the country level. Cost estimates usually emerge as part of a broader planning exercise to develop a country-level adaptation strategy. Early examples include studies sponsored by the World Bank in Bangladesh (Smith et al., 1998) and the Pacific (World Bank, 2000). A somewhat broader exercise is World Bank (2008), a recently initiated series of case studies that will deepen our understanding of adaptation costs in developing countries.

More significantly in terms of actual adaptation planning, the international community has embarked on a series of adaptation studies for the most vulnerable countries of the world, called the National Adaptation Programmes of Action (NAPAs). Around 40 NAPAs have so far been completed (see http://unfccc.int/adaptation/napas /items/4583.php). The aim of the NAPAs is to identify priority adaptations and initiate a process of planning, preparation and implementation in vulnerable developing countries.

NAPAs vary in quality and scope, with cost estimates ranging from less than $4 million in Madagascar, Comoros and the Central African Republic, to several hundred million dollars in Ethiopia and The Gambia, the only two countries to include extensive infrastructure investments. Elsewhere, NAPA priorities predominantly cover preparatory measures and capacity building, mostly on agriculture and water. As such, NAPAs are a poor indicator of the ultimate adaptation expenditures in vulnerable countries, although they can give a rough indication of what the initial outlay (and sectoral priorities) may be as the global adaptation effort is ramped up.

1.5 First-generation global estimates

Interest in global adaptation cost estimates increased sharply a few years ago when international support for adaptation in developing countries emerged as a key aspect in the negotiations for a successor to the Kyoto Protocol. In response to this demand a handful of aggregate adaptation cost estimates emerged in quick succession (Table 1.2). Although they are often dubbed 'global' they in fact concern adaptation in only the developing world.

Table 1.2 Estimates of adaptation costs in developing countries

	Cost (US$ billion per year)	Comments
World Bank (2006)	9 – 41	Cost of climate-proofing FDI, GDI and ODA flows
Stern (2007)	4 – 37	Update, with slight modification of World Bank (2006)
Oxfam (2007)	>50	Based on World Bank, plus extrapolation of costs from NAPAs and NGO projects
UNDP (2007)	86-109	World Bank, plus costing of PRS targets, better disaster response

Source: Agrawala and Fankhauser (2008)

Note: FDI = foreign direct investment, GDI = gross domestic investment, ODA = official development assistance, NAPA = National Adaptation Programme of Action, PRS = poverty reduction strategy

Another important feature of these estimates is that they are not independent. They are all based on the same method, first developed by the World Bank (2006). The Bank estimated the fraction of current investment flows that is climate sensitive and then used a 'mark up' factor that reflects the cost of 'climate-proofing' these investments. The Bank team assumed that 2–10% of gross domestic investment (GDI, worth $1500 billion a year at the time), 10% of foreign direct investment (FDI, $160 billion) and 40% of official development assistance (ODA, $100 billion) would be sensitive to climate change. The assumed mark-up to climate-proof these investments was 10–20%. Of these assumptions only the ODA figure had any empirical grounding. It was derived from earlier OECD work (Agrawala, 2005) about climate risks in six developing countries (Bangladesh, Egypt, Fiji, Nepal, Tanzania and Uruguay). The range found in that study was considerably wider though (12–65%), and its authors cautioned against drawing general conclusions.

Nevertheless, subsequent studies all adopted the same approach. The Stern Review (Stern, 2007) reduced the mark-up for climate-proofing from 10–20% to 5–20%, and the share of climate-sensitive ODA from 40% to 20%, but made no further adjustments to the method. The changes in assumptions were not explained, other than to say that they were derived 'through discussions with the World Bank'.

1

Oxfam (2007) adopted the World Bank numbers but added additional cost items, such as the extra cost of NGO work at community level and the cost of implementing a NAPA-style programme. Both cost items are based on fairly strong assumptions. The costs of community-level adaptation were extrapolated from just three projects, while the cost of early adaptation was derived from the 13 NAPAs available at the time.

The 2007 Human Development Report (HDR) (UNDP, 2007) again adopted the World Bank approach to costing infrastructure adaptation. The HDR used later, considerably higher investment data and a different share for climate-sensitive ODA (17–33%) but otherwise adopted the Stern assumptions. In addition, it included the costs of adapting poverty reduction strategies ($44 billion a year) and strengthening disaster response systems ($2 billion a year).

Although based on a common method, the studies resulted in a large range in estimates, with the lowest number at $4 billion and the highest one at over $100 billion. This indicates a fundamental problem with this estimation approach. There is little empirical information about the share of climate-sensitive investments and the mark-ups required to 'climate-proof' them, which are likely to be situation-specific. Yet, because investment flows are so large, even small changes in this uncertain parameter can change results by up to an order of magnitude.

1.6 The UNFCCC estimates

Perhaps the best global adaptation cost estimate available to date is UNFCCC (2007).[5] The Secretariat of the Framework Convention commissioned six sector studies to get a better idea of adaptation costs both globally and in developing countries. The estimates were made for the year 2030, usually assuming SRES A1B and B1 or similar scenarios.

- **Agriculture, forestry and fisheries.** The agriculture estimate consists of three distinct cost items: extra capital investment along the production chain (farms, transport, processing, etc.), the need for better extension services at country level and the cost of additional global research (e.g. on new cultivars). (See McCarl, 2007.)

- **Water supply.** The water estimate considers the effect of additional water demand and changes on the supply side. Investment decisions are made in anticipation of 2050 water needs. (See Kirshen, 2007.)

- **Human health.** The health estimate includes the extra treatment costs for three health issues: malnutrition, malaria and diarrhoea. Scenarios are based on the Global Burden of Disease study (McMichael et al., 2004). (See Ebi, 2007.)

- **Coastal zones.** Coastal protection costs are based on the DIVA model, which considers a limited set of adaptation options, which are applied globally. Uniquely, the coastal estimate considers both adaptation costs and residual damages. Investments are made in anticipation of 2080 sea level rise. (See Nicholls, 2007.)

- **Infrastructure.** The infrastructure estimate adopts the World Bank (2006) methodology, using insurance data to determine the share of climate-sensitive new investment. (See Satterthwaite, 2007.)

- **Ecosystems.** In the absence of firm data, an indication of adaptation costs for ecosystems was derived from the costs of increasing protected areas by 10%, although it was not possible to split this into baseline cost (e.g. prevention of land conversion) and incremental adaptation. (See Berry, 2007.)

UNFCCC (2007) concluded that total adaptation costs by 2030 could amount to $49–171 billion per annum globally, of which $27–66 billion would accrue in developing countries (Table 1.3). By far the largest cost item is new infrastructure investment, which for the upper-bound estimate accounts for three-quarters of total costs. Costs are over and above what would have to be invested in the baseline to renew the capital stock and accommodate growth. Note that the total excludes the estimate for ecosystem adaptation, which was not considered to be sufficiently robust.

Table 1.3 UNFCCC estimates of global investment costs for adaptation

Sector	Global cost ($bn per annum)	Developed countries	Developing countries	Residual damage
Agriculture	14	7	7	–
Water	11	2	9	–
Human health	5	0	5	–
Coastal zones	11	7	4	1.5
Infrastructure	8 – 130	6 – 88	2 – 41	–
Total	**49 – 171**	**22 – 105**	**27 – 66**	**1.5**

Source: UNFCCC (2007)

1.7 The state of the art

Most authors readily admit that adaptation cost estimates are preliminary, incomplete and subject to a number of caveats. A recent survey of cost estimates by the OECD came to the same conclusion (Agrawala and Fankhauser, 2008). Important gaps remain in terms of:

- the scope of the analysis (whether all relevant impacts and countries are covered)
- the depth of analysis (whether, for a given impact/country, all relevant adaptation options and needs are considered)
- the costing of measures (whether all relevant costs are included)
- the treatment of uncertainty (how uncertainty about future change affects costs).

1

Many of these gaps are exacerbated by the lack of a clear operational definition of adaptation. The question then is to what extent the prevailing knowledge gaps lead to an upward or a downward bias in the existing estimates. The sign of the bias is not immediately obvious, since there are omissions in both directions. However, it is likely that adaptation costs have been under-estimated so far (Table 1.4).

Table 1.4 Biases and omissions in existing cost estimates

Shortcoming	Likely bias
• *Scope*: Limited range of impacts considered	– – –
• *Depth*: focus on hard adaptation, no cost-effectiveness test	++
• *Depth*: focus on planned, public adaptation	– –
• *Costing*: ignore preparation, lifetime costs	–
• *Costing*: ignore higher-order effects	–
• *Uncertainty*: ignore need to plan for a range of outcomes	?

Source: author's own assessment

The main downward bias in existing studies has to do with their limited *scope*. Adaptation analysis is still predominantly a domain for case studies. Only a handful of studies aspire to provide a comprehensive global estimate, and the first generation of global studies (see Table 1.2 above) measured adaptation costs in only one particular sector, the cost of climate-proofing new investments.

Even the most extensive study to date (UNFCCC, 2007) is limited to six sectors, of which five were ultimately reported (agriculture, water, health, coastal zones, infrastructure). Some areas with clear adaptation needs – such as energy and tourism – were omitted, as were some prominent adaptation strategies that are likely to feature prominently, most notably migration.[6] Moreover, in the areas that were considered, the analysis was not always comprehensive. For example, only three health impacts were assessed (malnutrition, malaria and diarrhoea) while the water estimate excludes the cost of flood control and the extra expenditures required to maintain water quality.

The effect of insufficient *depth* is more difficult to assess. Even in detailed case studies it is difficult to consider systematically all available adaptation options. Adaptation is too broad and 'nebulous' (Agrawala and Fankhauser, 2008) a concept to capture analytically in a comprehensive way. The problem is exacerbated in global studies, which have to simplify and ignore local circumstances. This introduces at least two biases, one leading to an under-estimation and the other to an over-estimation of true adaptation costs.

The first bias is a preference for 'hard' structural adaptation measures over 'soft' behavioural or regulatory adaptations (agriculture is an exception). Hard adaptation measures, such as the expansion of water supply systems, are relatively easy to capture and generalise, but they are also potentially much more expensive than soft measures like changes in water demand (e.g., in response to price incentives). More generally, the proposed adaptation options are rarely subjected to a rigorous cost-effectiveness test, suggesting that other, more effective options might be identified once concrete options are being considered.

The second bias concerns the focus on public adaptation over private adaptation. This is again primarily because public adaptations are easier to identify than are the plethora of autonomous adaptations individuals and firms are likely to undertake. However, certainly in terms of quantity, and perhaps also in terms of costs, private autonomous measures will dominate the adaptation response as people adjust their buildings, change space-cooling and -heating preferences, reduce water use, alter holiday destinations or even relocate. UNFCCC (2007) includes many measures that will ultimately be the responsibility of private actors, in particular in health, agriculture and to some extent infrastructure, but it is likely that this represents only the tip of the iceberg in terms of private adaptation. Global cost estimates, including that of the UNFCCC, are primarily about planned adaptation.

Although *costing* is almost always ad hoc, the resulting bias is more difficult to ascertain. A careful assessment of adaptation costs would ideally consider the net present value of costs over the entire lifetime of a project, including preparation costs, investment costs, operations and maintenance costs and decommissioning costs. In reality, the focus is often on the initial capital expenditures (in the case of UNFCCC (2007), the focus is deliberately on investment costs). Another potential omission, not least in a developing country context, is institutional and administrative costs, including the costs of building planning capacity. There is some speculation, but little analytical evidence, about the importance of higher-order costs and inter-linkages, for example the effect of a large-scale coastal defence programme on construction activity elsewhere or the interaction between adaptation in water and agriculture.[7]

Finally, all cost estimates assume that future climate change impacts are known with *certainty*. In reality, adaptation planners will have to deal with considerable uncertainty about the likely extent and even direction of change at the local level. In some (rare) cases, this may justify a delay in adaptation until more information is available (thus reducing adaptation costs, but increasing residual impacts). In other cases, it will force planners to extend the scope of adaptations so that they can deal with a wider range of outcomes. This is likely to increase adaptation costs.

Readers interested in the overall costs of climate change will also have to remember that the scope of adaptation estimates is deliberately narrow. Important costs that matter for this broader debate were excluded to keep the problem tractable, including the costs of closing prevailing adaptation gaps, the cost of residual impacts and the cost of mitigation. Many of these will be higher than the cost of adaptation. Moreover, if the mitigation and development investments are not made, the cost of adaptation will be considerably higher.

1 More complex frameworks would also consider reasons for concern other than aggregate costs, such as the unfair distribution of impacts, the risk of tipping points, excessive climate variability and the threat to unique natural systems (Smith et al., 2001).

2 Chapter 8 in this report shows how the estimates of adaptation costs reviewed here translate into estimates of the social costs of carbon, that is the cost emissions impose on society over their atmospheric lifetime.

3 Smith and Tirpak (1989), Cline (1992), Nordhaus (1994), Fankhauser (1995), Pearce et al. (1995) and Tol (1995). A recent survey is Tol (2005).

4 For example, Nicholls and Tol (2006) on coastal zones and Parry et al. (2004) on agriculture. See Parry et al. (2007) for an assessment.

5 More accurately, the UNFCCC aimed to estimate the investment and financial flows for adaptation (and mitigation).

6 Planned migration away from affected areas can be an effective adaptation strategy. In contrast, the plight of climate refugees may be better classified as a residual damage as it reflects insufficient anticipatory adaptation. In addition, migration (e.g. a growing urbanisation) is also part of 'baseline' development.

7 See Bosello et al. (2007) for a rare general equilibrium study.

References

Agrawala, S. (2005). *Bridge over Troubled Waters: Linking Climate Change and Development*. Paris: OECD.

Agrawala, S. and S. Fankhauser (eds) (2008). *Economic Aspects of Adaptation to Climate Change. Costs, Benefits and Policy Instruments*. Paris: OECD.

Berry, P. (2007). *Adaptation Options on Natural Ecosystems*. A report to the UNFCCC Financial and Technical Support Division, Bonn.

Bosello, F., R. Ronson and R.S.J. Tol (2007). 'Economy-wide Estimates of the Implications of Climate Change: Sea Level Rise', *Environmental and Resource Economics* 37: 549–571.

Bowen, A., B. Chatterjee and S. Fankhauser (2009). *Adaptation and Economic Growth*. Paper prepared for the UK Department for International Development, Grantham Research Institute, London School of Economics, February.

Cline, W.R. (1992). *The Economics of Global Warming*. Washington DC: Institute for International Economics.

Ebi, K. (2007). *Health Impacts of Climate Change*. A report to the UNFCCC Financial and Technical Support Division, http://unfccc.int/cooperation_and_support/financial_mechanism/financial_mechanism_gef/items/4054.php.

Fankhauser, S. (1995). *Valuing Climate Change. The Economics of the Greenhouse*. London: Earthscan.

Kirshen, P. (2007). *Adaptation Options and Costs in Water Supply*. A report to the UNFCCC Financial and Technical Support Division, http://unfccc.int/cooperation_and_support/financial_mechanism/financial_mechanism_gef/items/4054.php.

McCarl, B. (2007). *Adaptation Options for Agriculture, Forestry and Fisheries*. A report to the UNFCCC Financial and Technical Support Division, http://unfccc.int/cooperation_and_support/financial_mechanism/financial_mechanism_gef/items/4054.php.

McGray, H., A. Hamill, R. Bradley, E.L. Schipper and J.-O. Parry (2007). *Weathering the Storm. Options for Framing Adaptation and Development*. Washington DC: World Resources Institute.

McMichael, A. et al. (2004). 'Climate Change', in: Ezzati, M., A. Lopez, A. Rodgers and C. Murray (eds), *Comparative Quantification of Health Risks: Global and Regional Burden of Disease*. Geneva: World Health Organization.

Nicholls, R. (2007). *Adaptation Options for Coastal Zones and Infrastructure*. A report to the UNFCCC Financial and Technical Support Division, http://unfccc.int/cooperation_and_support/financial_mechanism/financial_mechanism_gef/items/4054.php.

Nicholls, R. and R.S.J. Tol (2006). 'Impacts and Responses to Sea Level Rise: A Global Analysis of the SRES Scenarios over the 21st Century, *Philosophical Transactions of the Royal Society* A 363: 1073–1095.

Nordhaus, W.D. (1994). *Managing the Global Commons. The Economics of Climate Change*. Cambridge MA: MIT Press.

Noy, I. (2009). 'The Macro-economic Consequences of Disaster', *Journal of Development Economics* 88: 221–231.

Oxfam (2007). *Adapting to Climate Change. What is Needed in Poor Countries and Who Should Pay?* Oxfam Briefing Paper 104.

Parry, M.L. et al. (2007). *Climate Change 2007. Impacts, Adaptation and Vulnerability. Working Group II Contribution to the Fourth Assessment Report of the Intergovernmental Panel on Climate Change*. Cambridge: CUP.

Parry, M.L. et al. (2004). 'Effects of Climate Change on Global Food Production under SRES Emissions and Socio-economic Scenarios', *Global Environmental Change* 14(1): 53–67.

Pearce, D.W. et al. (1996). 'The Social Costs of Climate Change: Greenhouse Damage and the Benefits of Control' in: IPCC, *Climate Change 1995. Economic and Social Dimensions of Climate Change. Contribution of Working Group III to the Second Assessment Report of the IPCC*. Cambridge: CUP.

Satterthwaite, D. (2007). *Adaptation Options for Infrastructure in Developing Countries*. A report to the UNFCCC Financial and Technical Support Division, http://unfccc.int/cooperation_and_support/financial_mechanism/financial_mechanism_gef/items/4054.php.

Schelling, T., (1992). 'Some Economics of Global Warming', *American Economic Review* 82(1): 1–14.

Smith, J.B. et al. (2001). 'Lines of Evidence for Vulnerability of Climate Change: A Synthesis', in: IPCC, *Climate Change: Impacts, Adaptation and Vulnerability. Contribution of Working Group II to the Third Assessment Report of the IPCC*. Cambridge: CUP.

Smith, J.B. and D. Tirpak (eds) (1989). *The Potential Effects of Global Climate Change on the United States*. Washington DC: US Environmental Protection Agency.

Smith, J.B, A. Rahman, M.Q. Mirza, G.J. Kenny and G.C. Sims (1998). *Considering Adaptation to Climate Change in the Sustainable Development of Bangladesh*. Washington DC: South Asia Region, World Bank.

Stern, N. (2007). *The Economics of Climate Change: The Stern Review*. Cambridge: CUP.

Tol, R.S.J. (2005). 'The Marginal Damage Costs of Carbon Dioxide Emissions: An Assessment of the Uncertainties', *Energy Policy* 33(16): 2064–2074.

Tol, R.S.J (1995). 'The Damage Costs of Climate change – Toward More Comprehensive Calculations', *Environmental and Resource Economics* 5: 353–374.

Tol, R.S.J, Fankhauser and J.B. Smith (1998). 'The Scope for Adaptation to Climate Change: What Can We Learn from the Impact Literature?', *Global Environmental Change* 8(2): 109–123.

UNDP (2007). *Human Development Report 2007/08*. New York: Palgrave McMillan.

UNFCCC (2007). *Investment and Financial Flows to Address Climate Change*. Bonn: Climate Change Secretariat.

World Bank (2008). *The Economics of Adaptation. Concept Note and Study Plan*, April. www.worldbank.org/environment/eacc

World Bank (2006). *Investment Framework for Clean Energy and Development*. Washington DC: World Bank.

World Bank (2000). *Cities, Seas, and Storms. Managing Change in Pacific Island Economies. Volume IV: Adapting to Climate Change*. Washington DC: Papua New Guinea and Pacific Island Country Unit, World Bank.

2

Costs of adaptation in agriculture, forestry and fisheries

Tim Wheeler and Richard Tiffin

2

Summary

Production and activity within the agriculture, forestry and fisheries sector is inherently affected by variability in climate. There is a tradition of coping with year-to-year changes in climate; nevertheless, human-induced climate change is expected to push these managed ecosystems beyond their natural climatic boundaries, requiring a greater rate and extent of adaptation than previously needed. In this chapter we examine the methods and assumptions of costing this adaptation to climate change within the agricultural, forestry and fisheries sector.

A range of methods exists for costing adaptation. The UNFCCC report *Adaptation Options for Agriculture, Forestry and Fisheries* (McCarl, 2007) takes a top-down approach to costing adaptation. A 10% increase in research and extension funding and a 2% increase in capital infrastructure costs are assumed due to climate change using the A1B1 SRES scenario for the year 2030, termed 'without mitigation'. The global marginal estimates for additional funding for adaptation of the agricultural sector due to climate are $12.6 billion and $11.3 billion – without and with mitigation, respectively, in the year 2030 (McCarl, 2007). The cost of efforts to keep up with the demands from the sector with future changes in population are given as more than an order of magnitude higher, at $520 billion. Alternative bottom-up approaches focus on costing specific adaptation options, such as improved irrigation or developing new adapted crop varieties. For example, interpretation of the study of Fischer et al. (2007) provides a cost of $8 billion for adapting crop irrigation systems to climate change by 2030. A different approach by Cline (2007), using simple crop growth models, provides an estimate of $14.5 billion for the year 2030 for the reduction in the value of global crop outputs due to climate change.

We conclude that estimating the cost of adaptation of the agriculture, forestry and fisheries sector to climate change is a difficult task requiring further research. The McCarl (2007) estimate from the UNFCCC takes a top-down approach that projects forward increases on current costs to provide a figure of $11.3–12.6 billion for the year 2030. We conclude that this is a reasonable first approximation of the additional costs of adaptation of agriculture, forestry and fisheries. Nevertheless, on the basis of the limited independent evidence available, we expect bottom-up approaches to reveal this to be on the low side of adaptation costs in this sector.

2.1 Assumptions

When measuring the economic cost of an activity it is important to understand the purpose for which the cost is being evaluated. The usual reason for measuring cost is to compare the costs and benefits of a project in order to judge whether it should be implemented. The issues associated with such an evaluation are complex, and include a recognition that market prices may not represent the true value of the project owing to distortions in the market, and the fact that many benefits and costs are not subject to market transactions and therefore do not have readily observed values. In many cases the cost-benefit analysis (CBA) is conducted from the perspective of a single economic agent. For example, a firm may conduct a CBA of a particular investment project, a government might analyse a scheme to protect a site of environmental value, and consumers might be interested in the costs and benefits of replacing their car. In a broader context, such as that under consideration in this chapter, the evaluation does not pertain to a single economic agent but to society as a whole. Moreover, the nature of the 'project' to be evaluated needs careful consideration.

Climate change is an exogenous shock to the economy and adaptation and mitigation are responses to that shock. There are qualitative differences between these responses. One aspect of these qualitative differences is considered by Callaway (2004), who argues that adaptation to climate change is essentially private, in contrast to mitigation. The consequences of the adaptation action can be excluded and therefore accrue primarily to their instigator. For example, the sole beneficiary of the decision by a farmer to adjust the planting date of his crop will be the farmer himself. The distinction is important, because the existence of a public good will lead to a market failure which in turn leads to an under-provision. In contrast, where a good is private, the market can be expected to provide an efficient outcome if there are no distortions (a big if). One reason for evaluating the costs of adaptation is to determine the optimal combination and level of adaptation and mitigation.

While viewing adaptation to climate change as a private good serves to make a point, it is probably an oversimplification. There will be aspects of adaptation that are also public goods, for example the research and development of new varieties adapted to new climates. The other important consideration relating to adaptation is its distributional consequences. While the market might solve the climate problem through adaptation, the distribution of the market benefits might be substantially different. For example, if prices rise as a result of the adaptation, producers may benefit at the expense of consumers. Moreover, the distribution of the benefits among producers may be spatially uneven with, for example, developed country farmers benefiting at the expense of those in developing countries.

The distinction between public and private goods is important because the presence of the former implies that the market will not provide a welfare-maximising response to the exogenous shock. As a consequence, government intervention is justified, to improve welfare. In this context, the cost that needs to be measured is that of the proposed government intervention, and this needs to be compared with the welfare loss that results from the shock as a measure of the benefit from intervention.

The perspective may not be one of market failure, however. One might be interested simply in how much money it is going to cost us to make the necessary adaptations to climate change. This is essentially the focus of the McCarl (2007) paper. In this context it is important to recognise that what we seek is a measure of the opportunity cost that is forgone as a result of us needing to adapt to climate change. For example, we might have to spend money on developing new varieties capable of withstanding drought stress. In the absence of climate change we might have equally spent money on developing varieties that are better adapted to the unchanged climate. What matters is the additional expenditure that is necessary to adapt. Once we have measured a cost of this sort it is important to recognise its significance. Spending money is not of itself a bad thing and one person's cost is another person's benefit. When we measure costs of this sort what we often measure are transfers from one group to another. This is not to say that distributional questions are not important. The problem is that the answer as to whether a move is good or bad is much harder as it depends on us making value judgements about the importance of different groups in society. From a social perspective, the only thing that is unambiguously bad is a market failure which results in an overall reduction in output or welfare.

2

The final conceptual issue which is particularly important in the context of climate change is the distinction between static and dynamic costs. In a static approach the cost that is measured is one that compares two discrete situations. In this case, one with climate change and one without. There are costs that are associated with making the change which may be ignored in a static approach. These are termed adjustment costs and are the focus of a dynamic approach. With climate change these costs may include things such as migration and reduction. A population could have an equally high level of welfare in a new location once it had established itself, but there are likely to be substantial reductions in welfare that are the consequence of having to relocate.

2.2 Issues

There are a wide range of possible adaptation strategies to climate change within agriculture, fisheries and forestry. These span a range of scales from farms to governments, and from relatively simple autonomous changes in the way agricultural businesses are managed to complex programmes of research and development lasting decades. Many authors provide lists and categories for adaptation strategies for this sector, with much in common from one to the other. The review by Howden et al. (2007) provides a good overview of adaptation in agriculture, forestry and fisheries. It considers many adaptation options as either changes in management decisions or changes in the decision environment. Changes in management within cropping systems include altering crops or crop varieties, more efficient management of water, altering the timing or location of cropping, and improving the effectiveness of crop protection measures. Within livestock systems many adaptation options are connected with maintaining the availability of fodder and feed and reducing heat stress from animal housing. Adaptation of managed forests could involve changes in tree species composition, harvesting patterns, pest control and location of managed woodland. Marine fisheries adaptation is less sensitive to management changes, except for changes in catch size (Howden et al., 2007). Adaptation of the sector through decision making may include policy changes, the development of infrastructure and general changes to the decision-making environment.

It is also clear that there is an enormous range of actors within the sector as well: individuals and collectives, private and public institutions. Given this diversity and complexity of possible adaptation options and actors, how well can the spectrum of possible adaptation strategies within agriculture, forestry and fisheries be costed and summarised in a single or a few global headline figures?

2.3 Existing methods

The UNFCCC report *Adaptation Options for Agriculture, Forestry and Fisheries* (McCarl, 2007) takes a top-down approach to costing adaptation. It splits adaptation costs for the agriculture, forestry and fishery sectors taken as a whole into those needed for research, extension and physical capital expenditure. It then projects forward the past trends in each of these, sourced from current estimates and the literature, and imposes an additional increase due to climate change. A 10% increase in research and extension funding and a 2% increase in capital infrastructure costs are assumed due to climate change using the A1B1 SRES scenario for 2030, termed 'without mitigation'. These additional costs are then reduced for the SRES B1 'mitigation' scenario by multiplying by the ratio of the projected global temperature change for A1B1 relative to B1; that is 1.4/1.6.

The global estimates for additional funding for adaptation of the agricultural sector due to climate change in the year 2030 are $12.6 billion and $11.3 billion, without and with mitigation respectively (McCarl, 2007). The cost of efforts to keep up with the demands from the sector with future changes in population are given as more than an order of magnitude higher than this at $520 billion.

2.4 Critique of methods

Although the approach of McCarl (2007) seems just to take the current state and add a more or less arbitrary amount to represent the additional costs of adaptation, it is very hard to judge whether the magnitude of the figure is reasonable without trying to take a bottom-up approach to costing. One useful outcome seems to be that there is already a substantial amount invested in research and development within the sector and that adaptation is already well embedded. As a consequence, it is clear that the incremental costs of adaptation are likely to be small (relatively speaking).

Can global estimates of the cost of adaptation in the agricultural sector be derived from the bottom up as an alternative approach to that taken by McCarl (2007)? A number of options are available. The first is to try explicitly to model impacts and different adaptation strategies and then to cost them. For example, much progress has been made in assessing the impacts of global climate change on crop production to date. Research is moving towards a consensus that predicts a small increase in production of the major grain crops across northern, developed nations up to 2050, followed by a gradual decline to 2100, and a steady decline in yields in tropical regions over this entire period (Parry et al., 2004; Cruz et al., 2007). The effects of climate change on world cereal prices follow these trends in crop yields, with little consistent change in output price among different studies until a warming of 2–3°C from current climates, beyond which prices start to rise uniformly across different studies. These approaches can be extended to consider upstream and downstream impacts on costs through the use of a general equilibrium model (for example, Winters et al., 1998; Lewandrowski and Schimmelpfennig, 1999).

Second, a meta-analysis of all existing impacts and adaptation studies can be brought together on a common axis as a surrogate for climate change. Such an exercise was undertaken by Easterling et al. (2007) who, from many maize, rice and wheat studies, quantified the impacts of and adaptation to a range of temperature warming on these crops. At mid- to high latitudes, where the majority of global cereal production is found, a moderate warming of 1–3°C (expected in the decades close to 2050) is thought to have a small beneficial effect on wheat yields of 5–10%, but warming above 3°C reduces yield below current values. At low latitudes, the yield of wheat declines steadily with any warming. Adaptation of cropping practices improves yield across the temperature warming range, and hence pushes the onset of negative impacts further into the future. For example, the potential benefits of adaptation of cropping practices to climate change may be as much as an 18% improvement in the yield of temperate and tropical wheat systems. The response of rice to temperature warming is broadly similar to that of wheat. For maize, there is less benefit to yield of small temperature changes at mid- to high latitudes and a more dramatic decline at warmer temperatures, even with adaptation. This meta-analysis is still one step away

2

from providing adaptation costs, but does sample a lot of empirical evidence and defines the response of adaptation to different magnitudes of climate change, as represented by global temperature changes.

The use of hedonic modelling presents a positive alternative to the normative approaches that are based on simulation through the use of integrated climate, crop and economic models. The use of hedonic equations assumes that the impacts of climate change on profitability will be capitalised into the value of land and yields. They carry the advantage of not being constrained by the adaptation strategies that are deemed feasible by the parameterisation of the models used in a simulation approach. An equation relating land values to climate variables is then estimated and used to forecast the impact of changes in these variables on land values over a number of different climate change scenarios. This approach is exemplified by Schlenker et al. (2005; 2006). Deschenes and Greenstone (2007) note that the hedonic approach is subject to criticism because unobserved factors which influence land values are excluded from the hedonic equation. As a result it is probable that the estimates which result from this approach are biased. They therefore propose an alternative approach in which profitability is directly related to climatic variables. They note that this may also produce biased results because it fails to take into account the full extent to which farms may adapt to climate change. They argue that because the direction of bias is known (an under-estimate of the impacts), their method is superior to the hedonic approach.

2.5 Comparison with case studies

Compared with McCarl (2007), the examples of bottom-up approaches are more closely linked to individual processes of adaptation within the agricultural sector. Can studies of the cost of single adaptation strategies, or on a local or regional scale, provide an independent assessment of the UNFCCC global cost? Potentially they could, through careful selection of representative sites, enterprises and adaptation options, but costed studies on the local or regional scale are also scarce in the literature.

Some studies provide costs of single adaptation options. For example, Fischer et al. (2007) estimated adaptation costs through meeting future irrigation demands by 2080 to be $24–27 billion per year for an unmitigated scenario and $8–10 billion per year less for the mitigated B1 scenario. These adaptation costs of providing increased irrigation capacity can be compared with the estimates of McCarl (2007) by bringing them to a common metric under the same emissions scenario, where this is possible. The total additional costs of McCarl (2007) are $12.6 billion and $11.3 billion of additional adaptation costs for the year 2030 for the A1B and B1 emissions scenarios, respectively. The unmitigated scenario used by Fisher et al. (2007) was different from that of McCarl (2007). Under the same mitigated B1 scenario the study of Fischer et al. (2007, Figure 3) gives additional annual costs of $8 billion per year for increasing irrigation capacity for 2030. These additional adaptation costs to maintain irrigation capacity under climate change are therefore about 65% of those of McCarl (2007) although we are comparing the cost of a single adaptation option in the crops sector to costs across agriculture, forestry and fisheries. Of course our confidence in such a comparison using this bottom-up approach to estimate the additional costs of adaptation to climate change are highly uncertain using the limited evidence currently available. Nevertheless, if we were to extend this approach with costings for other explicit adaptation measures, it could be that the additional adaptation costs for the agricultural sector will exceed the estimates of McCarl (2007).

The cost of developing new crop varieties in public breeding programmes has been put at $2.1 million for oats and $2.8–3.0 million for wheat for a single variety (NRC, 2000). The Alliance for a Green Revolution in Africa puts the cost of developing 200 new crop varieties better adapted to local environments at $43 million. The development of Bt maize by Monsanto is thought to have cost $10–25 million (NRC, 2000). These examples for developing new crop varieties show a wide spread of costs for at least some components of global estimates for the sector, but are all tiny in comparison with the total estimate for the sector of McCarl (2007). It is very difficult to see how this bridge across the divide of spatial scale and process detail to a global scale could be made. Perhaps it cannot, so that we should view the global estimates of adaptation cost together with (unconnected) enterprise- or location-specific studies. It is also difficult to separate population-related needs from needs related to climate change.

Within the economics literature two approaches that can be interpreted as the net costs of adaptation exist for costing the impacts of climate change. These are referred to as the Ricardian and the crop growth model methods. In the former, land values are modelled under an assumption that they reflect the future profitability of farms (Mendelsohn et al., 1994). Studies employing the Ricardian approach are country specific and show a diverse array of effects, with some countries (e.g. the US) benefiting and others losing (Sanghi and Mendelsohn, 2008). In the crop growth model approach, the impacts of climate change are simulated using crop growth models and the value of the resulting change in output is taken as a measure of the economic impact of climate change. Cline (2007) summarises these estimates for in excess of 70 countries. He estimates that the overall impact of climate change on agriculture will be a reduction in the value of output of the order of $38 billion dollars by 2080. This figure masks some wide variations in the spatial impacts, however. For example, India alone is predicted to witness a reduction in its output of the value of $38 billion dollars, which is offset by gains elsewhere including China ($14 billion) and the US ($8 billion). However, if we simply scale back this estimate to the year 2030, assuming (probably incorrectly) a linear increase in cost over time, the Cline (2007) estimate equates to $14.3 billion for the year 2030. This is very close to the estimate of McCarl (2007). Nevertheless we need to bear in mind that the Cline (2007) study is for adaptation of crop production alone, while McCarl (2007) is for the entire agriculture, forestry and fisheries sector.

2.6 Outstanding issues

We have argued above that the McCarl approach can be criticised from a theoretical perspective as a result of being insufficiently clear regarding the objectives of measuring the cost. In this section we examine some more detailed aspects of this approach.

Costs of research in the UNFCCC report are based on information from the Consultative Group on International Agricultural Research (CGIAR) institutes. These omit national and private-sector research efforts, both of which may have a different magnitude of costs and cost-benefit ratios, particularly for agricultural research and development (R&D) in developed countries. However, there is a close relationship between increases in agricultural productivity and investment in R&D (World Bank, 2008), with an average rate of return on investment in agriculture R&D found to be 43% across 1673 studies worldwide.

2

As the variability of climate increases, farmers and agribusiness will have to adapt to different patterns of rainfall and more extremes of rainfall and temperature than currently experienced. Variability in climate can have important and dramatic impacts on the productivity of cropping systems (Porter and Gawith, 1999; Wheeler et al., 2000). Conceptually, we could see this as a challenge different from that of adapting to changes in mean climate conditions. To adapt to increased climate variability, those managing farms, forests and fisheries will have to adjust their risk-management strategies to try to maintain production and profits or, for some, to counter their losses or better exploit good years. Deciding how to cost in what is effectively a probabilistic adaptation response is difficult, but nevertheless important. It is likely to take a form quite different from a simple linear enhancement of spending on adaptation. At a conceptual level, the cost associated with increased variability should be measured as the additional resources that society expends in avoiding the associated risk. In a practical sense this is perhaps relatively straightforward and may, for example, be a matter of measuring the incremental cost of fertiliser, irrigation and crop protection. Conceptually the cost is determined by the size of the increase in variability and the attitude of the producer to risk. The issue is further complicated by the recognition that the costs are likely to spill over into sectors of the economy beyond the primary producers. Intermediaries and final consumers are likely to adopt strategies designed to mitigate the impacts of risk on their profitability and welfare.

The development of agriculture occurs against a moving background of changing demand for food and agricultural products, each with their own features specific to different locations, different rates of change and timescales of responses. McCarl (2007) provides a useful contrast of the magnitude of the costs of providing for an expanding population that is about 40 times greater than the cost of adapting to climate change. This suggests that adaptation/development costs within the sector will be dominated by changes driven not directly by climate. Nonetheless, there are many intrinsic links between increases in productivity and resilience to climate that confuse this discrimination as a basis for costing.

A clear and consistent conclusion from several decades of studies of the impacts of climate change on agriculture is the uneven distribution of positive and negative impacts in different parts of the world. The precise balance between positive and negative impacts depends on the type of agricultural business and the degree of climate change, but can roughly be summarised as positive impacts and opportunities for agriculture at the higher latitudes over the coming decades in contrast to more immediate negative impacts at lower latitudes. Thus, a global cost of adaptation needs to account for changes in the balance of benefits or opportunities and negative impacts in different parts of the world and over a range of time projections – a complex situation. Nevertheless, in the agricultural sector it is important to account for opportunities in some regions where production may potentially benefit from a moderate amount of climate change. Here, farmers could exploit short-term opportunities for their businesses with some investments costs, but with little or no investment would miss these. Although such investments are unlikely to be seen as adaptation to climate change, they could dominate spending in the short term in some regions. In addition, there is the potential for adaptation funds to be directed to developing countries within any global total. However, to date such adaptation funds have been comparatively small, for example, $150–300 million per year (World Bank, 2008).

Interaction between the degree of climate change, how sensitive agriculture in different parts of the world is to climate, and the level of farming technology available locally will be important in determining adaptive capacity. Part of this complex interaction is accounted for by considering, as the UNFCCC report does, developed and developing nations separately. This is sensible, but there are obvious exceptions. It could be that the exceptions outnumber the general rule once further divisions are considered, thus providing more room for error.

2.7 Estimate of residual damage

Costs of the residual damage from climate change after adaptation of agriculture, forestry and fisheries are difficult to find. An example estimate of the residual damage from adaptation of agriculture can be provided by interpretation of the adaptation study of the wheat crop in Australia. The effect of adaptation through changes in sowing dates and crop varieties on the gross value of the national wheat crop was simulated by Howden and Jones (2004). Uncertainty in response was represented using an ensemble of IPCC scenarios and spatial variability across Australia. Without adaptation the maximum potential increase under climate change for 2080 was 10% of gross value of the national crop – $0.3 billion per year (Figure 2.1). With adaptation there was a median increase of $158 million per year in crop value, but with a range of values about this, many below zero, and a small chance of a 20% decrease in crop value ($0.6 billion per year) with the uncertainty that was sampled (Figure 2.1). As a proportion, this represents a residual damage of about 20%.

Figure 2.1 Changes in national gross values of wheat in Australia for 2070
(a) without adaptation; (b) with adaptation

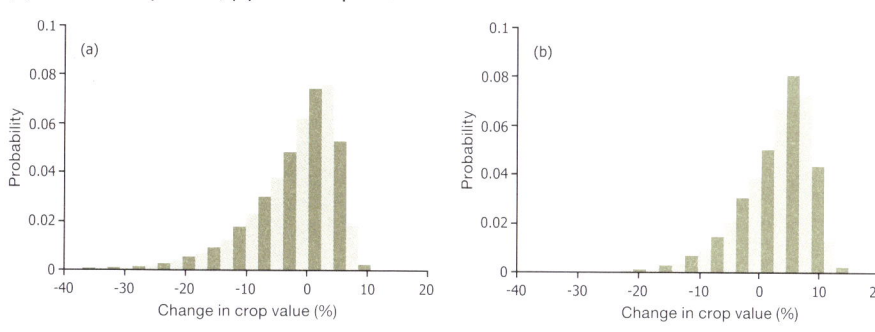

Source: Howden and Jones (2004)

2

2.8 Conclusions

We have concentrated in this paper on why it is difficult to estimate a single global cost for the adaptation of agriculture, forestry and fisheries to climate change. The estimate of McCarl (2007) is the only example of a global estimate specifically for this sector. However, we argue that this approach is certainly difficult to verify. A bottom-up comparison with this global figure using single adaptation options provides orders of magnitude smaller than (crop variety) or more than half of the total of (irrigation) this global value for the entire sector. Whilst not providing a direct link to the global figure, these single components of adaptation do suggest that $12.6 billion and $11.3 billion without and with mitigation could be an under-estimate of the cost of global adaptation of agriculture, fisheries and forestry.

Our interpretation of the crop growth model study of Cline (2007) is in good agreement with the UNFCCC estimate of McCarl (2007), with the important difference that Cline (2007) studied crop production alone whilst McCarl (2007) covered agriculture, forestry and fisheries. Therefore, we conclude that the UNFCCC estimate of $11.3-12.6 billion is a reasonable first approximation of adaptation costs in this sector, but we expect the estimate of adaptation costs for agriculture, forestry and fisheries to increase as more detailed studies of specific adaptation actions become available and as our understanding of the impacts of climate change matures.

References

Callaway, JM (2004) 'Adaptation Benefits and Costs: Are they Important in the Global Policy Picture and How Can We Estimate Them?', *Global Environmental Change* 14: 273–282.

Cline, WR (2007) *Global Warming and Agriculture: Estimates by Country*. Washington: Center for Global Development and Peterson Institute for International Economics.

Cruz, RV, Harasawa, H, Lal, M, Wu, S, Anokhin, Y, Punsalmaa, B, Honda, B, Jafari, M, Li, C and Ninh, NH (2007) *Climate Change 2007: Impacts, Adaptation and Vulnerability: Contribution of Working Group II to the Fourth Assessment Report of the Intergovernmental Panel on Climate Change*, ML Parry, OF Canziani, JP Palutikof, PJ van der Linden and CE Hanson (eds). Cambridge: Cambridge University Press, 469–506.

Deschenes, O and Greenstone, M (2007) 'The Economic Impacts of Climate Change: Evidence from Agricultural Output and Random Fluctuations in Weather', *American Economic Review* 97: 354–385.

Easterling, WE, et al. (2007) 'Food, Fibre and Forest Products', *Climate Change 2007: Impacts, Adaptation and Vulnerability: Contribution of Working Group II to the Fourth Assessment Report of the Intergovernmental Panel on Climate Change*, ML Parry, OF Canziani, JP Palutikof, PJ van der Linden and CE Hanson (eds). Cambridge: Cambridge University Press, 273–313.

Fischer, G, Tubiello, FN, van-Velthuizen, H and Wiberg, DA (2007) 'Climate Change Impacts on Irrigation Water Requirements: Effects of Mitigation 1990–2080', *Technological Forecasting and Social Change* 74: 1083–1107.

Howden and Jones (2004), cited in Hennessy, KB, Fitzharris, BC, Mates, N, Harvey, SM, Howden, L, Hughes, L, Sainger, J and Warrick, R (2007) 'Australia and New Zealand', *Climate Change 2007: Impacts, Adaptation and Vulnerability: Contribution of Working Group II to the Fourth Assessment Report of the Intergovernmental Panel on Climate Change*, ML Parry, OF Canziani, JP Palutikof, PJ van der Linden and CE Hanson (eds). Cambridge: Cambridge University Press, 507–540.

Howden, SM, Soussana, JF, Tubiello, FN, Chhetri, N, Dunlop, M and Meinke, H (2007) 'Adapting Agriculture to Climate Change', *Proceedings of the National Academy of Sciences* 104: 19691–19696.

Lewandrowski, J and Schimmelpfennig, D (1999) 'Economic Implications of Climate Change for US Agriculture: Assessing Recent Evidence', *Land Economics* 75: 39–57.

McCarl, BA (2007) *Adaptation Options for Agriculture, Forestry and Fisheries. A Report to the UNFCCC Secretariat Financial and Technical Support Division*.

Mendelsohn, R, Nordhaus, WD and Shaw, D (1994) 'The Impact of Global Warming on Agriculture: A Ricardian Analysis', *American Economic Review* 84(4): 753–771.

NRC (National Research Council) (2000) *Genetically Modified Pest-protected Plants*. Washington: National Academy Press.

Parry, ML, Rosenzweig, C, Iglesias, A, Livermore, M and Fischer, G (2004), 'Effects of Climate Change on Global Food Production under SRES Emissions and Socio-economic Scenarios', *Global Environmental Change* 14: 53–67.

Porter, JR and Gawith, M (1999) 'Temperatures and the Growth and Development of Wheat: A Review', *European Journal of Agronomy* 10: 23–36.

Sanghi, A and Mendelsohn, R (2008) 'The Impacts of Global Warming on Farmers in Brazil and India', *Global Environmental Change* 18: 655–665.

Schlenker, W, Hanemann, WM and Fisher, AC (2005) 'Will US Agriculture Really Benefit from Global Warming? Accounting for Irrigation in the Hedonic Approach', *American Economic Review* 95: 395–406.

Schlenker, W, Hanemann, WM and Fisher, AC (2006) 'The Impact of Global Warming on US Agriculture: An Econometric Analysis of Optimal Growing Conditions', *Review of Economics and Statistics* 88: 113–125.

Wheeler, TR, Craufurd, PQ, Ellis, RH, Porter, JR and Vara Prasad, PV (2000) 'Temperature Variability and the Yield of Annual Crops', *Agriculture, Ecosystems and Environment* 82: 159–167.

Winters, P, Murgai, R, Sadoulet, E, De Janvry, A and Frisvold, G (1998) 'Economic and Welfare Impacts of Climate Change on Developing Countries', *Environmental and Resource Economics* 12: 1–24.

World Bank (2008) *World Development Report; Agriculture for Development*. Washington DC: The World Bank.

3

Costs of adaptation in the water sector

Nigel Arnell

3

3.1 Introduction

One of the largest impacts of climate change is likely to be on water resources and their management. While there have been many studies of hydrological impacts at catchment, regional and global scales, there have been very few published studies on the costs of potential adaptation options at any scale.

The UNFCCC report (UNFCCC, 2007) provides the first global-scale estimates of the potential costs of adaptation. The headline conclusion is that the additional investment and financial flows required for adaptation to potential changes in the availability of water supplies would be approximately $9–11 billion per year in 2030, approximately 85% of which would be needed in non-Annex 1 Parties.[1] The UNFCCC report suggests that this is the same order of magnitude as the additional investment required to meet the Millennium Development Goal targets for sustainable access to safe drinking water and basic sanitation.

However, there are grounds (outlined below) for expecting that the UNFCCC figure is an underestimate of likely potential costs of adaptation to water supply. The figure does not include an allowance for costs of adapting in other aspects of water resources management, such as managing increased flood risk, maintaining water quality standards and supporting instream economic and environmental uses. The figure does not include any additional costs of residual impacts of climate change, caused by events above the design standard, and these costs may be very substantial in practice because adaptation will be neither perfect nor completed on time. Also, the figure of $9–11 billion/year represents the *additional* cost of maintaining a defined level of service in the presence of climate change. It does not include the cost of providing this level of service where it does not currently exist. The UNFCCC report estimates that the total cost (to 2030) of providing a specific level of service globally – thus reducing the development deficit – varies between $639 billion and $797 billion, depending on assumed economic and population growth, corresponding to an annual average of approximately $32–40 billion (assuming investment is over a 20-year period).

This chapter provides a critique of the UNFCCC estimates, but first examines the conceptual and methodological issues around the estimation of adaptation costs and residual damages in the water sector. It also reviews the (limited) literature.

3.2 Issues and assumptions

The 'water sector' is very diverse, and climate change is anticipated to impact upon many activities. These activities include: the supply of safe water to domestic, industrial and agricultural consumers (including for irrigation); the provision of sanitation and the removal and treatment of effluent; support of navigation; management of flood hazard (from drains, rivers, groundwater, overland flow and so on); measures to provide protection or reduce exposure; generation of hydropower; and the management of river flows and water levels to support agriculture, recreation and the provision of ecosystem services (such as support for instream and riverine ecosystems).

Estimates of the total current expenditure on water infrastructure are uncertain; one estimate (Briscoe, 1999) gives a total annual expenditure of $65 billion *in developing countries* for

3

hydropower ($15 billion), water supply and sanitation ($25 billion) and irrigation and drainage ($25 billion), but this excludes flood management costs. Table 3.1 summarises the potential economic and non-economic consequences of the effect of climate change on the water sector, assuming no explicit adaptive response to climate change.

Table 3.1 Potential consequences of climate change on the water sector, without adaptation

Activity	Potential economic consequences	Potential non-economic consequences
Water supply		
Domestic/ municipal	Cost of altered health Cost of dealing with droughts	Disruption to established patterns of use
Industrial	Cost of change in industrial productivity	Disruption and uncertainty
Agricultural (including irrigation)	Cost of change in agricultural productivity	Uncertainty
Sanitation and effluent removal	Cost of altered health Cost of impacts on instream ecosystems (e.g. fisheries) Cost of dealing with pollution incidents	
Navigation	Cost of altered navigation opportunities	
Flood management	Change in economic value of flood damages (direct and indirect) Change in economic value of injury and ill health	Disruption and anxiety
Hydropower	Cost of change in generation potential	
Recreation	Cost of changes in recreational opportunities	Change to cultural value of the water environment
Water level and soil water management		Change in habitat characteristics
Ecosystem services		Change in instream and riverine habitats and species

42 | Assessing the costs of adaptation to climate change

3

In practice, of course, water users and managers will adapt to a changing climate – possibly inappropriately and probably reactively – and so the actual cost of climate change will not be equal to the simple costs of impacts outlined in Table 3.1. For example, water supply companies are likely to introduce a variety of measures to try to maintain the security of supplies to customers, and consumers exposed to water shortage will seek alternative sources of supply. Managers of a wetland will alter pumping regimes to seek to maintain target water levels.

Measures taken to respond to other pressures on the water environment will also affect the impacts of climate change. If these other pressures and the responses to them are ignored for the moment, then the 'cost' of climate change for a given activity in the water sector is:

Cost = cost of explicit adaptation measures
 + residual impacts of climate change
 + transaction costs of implementing adaptation measures

The cost of explicit adaptation measures here represents the cost of measures explicitly introduced, by all interested parties (consumers and providers), to cope with climate change, and does not include the cost of measures introduced to meet other challenges which incidentally help adaptation to climate change. In practice, it is likely that in many cases adaptation measures will be developed to meet a variety of challenges, and it may be difficult to separate out the portion associated explicitly with climate change.

Adaptation will not remove all the consequences of climate change, and there will be residual impacts. These impacts will arise because adaptation may lag behind the changing climate, or because the adaptation measures introduced do not cope with the change in climate that actually occurs (due to imperfect knowledge). Impacts will also occur, as in the absence of climate change, in events which exceed the design standards or provisions of the adaptation measure; the net effect of climate change is the difference between such residual impacts with and without climate change. These residual impacts will be difficult to quantify, as they vary across different parts of the water sector. They are relatively easy to characterise in flood management – they are the costs incurred in damaging events which are not prevented – but in other parts of the sector the residual impacts take the form of lost productivity (agriculture, industry and power generation), inconvenience and exposure to water-related disease and ill health. Residual impacts also include the cost of emergency actions taken in response to an event, including the cost of implementing drought management plans and measures.

Transaction costs are the costs associated with making changes to policies and practices in the face of potential climate change. They include R&D costs and the costs of refining policies or reviewing decisions; these costs will be incurred even if decisions are subsequently made not to adapt to climate change.

The objectives of adaptation will influence the costs of adaptation and the residual impacts. In the most general terms, adaptation can aim at: (i) maintaining a given standard of service, (ii) achieving a new 'optimum' standard of service, or (iii) meeting some new service standard. This new service standard could be higher – because for example the threat of climate change increases risk aversion – or could be lower because of financial or feasibility constraints. These objectives

3

will vary from place to place and activity to activity. In many parts of the water sector, standards are set by some form of regulation (design standards for flood defence, for example, or reliability standards for the provision of drinking water or the discharge of effluent), but in other parts 'levels of service' are determined locally.

In practice, many parts of the world do not enjoy 'acceptable' standards of service, even in the absence of climate change. For example, a billion people currently lack access to safe drinking water, and 2.4 billion lack access to basic sanitation. A key conceptual question in assessing the costs of adaptation is therefore: should the costs of adaptation to climate change be (i) the cost of maintaining current standards of service, even where they are currently inadequate, or (ii) the cost of providing a defined standard of service? Under the first rule, costs of adaptation would be estimated to be extremely low where, for various historical reasons, standards of service are low (the 'development deficit' is high); under the second rule, the costs could be very high. The resolution of this conundrum requires an evaluation of ethical issues associated with the adaptation deficit and historical responsibilities both for climate change and for economic development.

3.3 Methodologies: approaches to costing

The previous section highlighted the issues associated with estimating the actual and potential costs of adaptation in the water sector. In principle, the equation above can be used both to characterise the actual (realised) cost of impacts and adaptation in the water sector, and to estimate in advance what these costs may be. In practice, of course, analysts are most interested in projections of future adaptation and impact costs.

It is relatively easy (in principle) to estimate future adaptation and residual damage costs at the water management system or scheme level. The simplest approach is first to estimate (over the next few decades) capital and operating costs (over all actors) plus residual damages, in the absence of climate change (but including other changes, such as the effects of changes in demand or exposure), and then repeat the calculations assuming a climate change trend and a management strategy specifically designed to cope with this trend. In circumstances where a development deficit is projected to persist, the reference 'no climate change' case can represent continued inadequate adaptation to climatic variability; alternatively, the reference case can include improvements to meet enhanced standards. The choice does not affect the estimation methodology.

This simplest approach assumes perfect knowledge: the water managers know exactly how climate will change, and plan in advance, keeping pace with the changing climate. A variation on this approach assumes 'perfect, but delayed' knowledge, whereby there is a lag between change in climate and the management response, with subsequent rescheduling of adaptation costs but an increase in residual impacts.

A more complicated approach recognises that water managers do not know how climate will change, and therefore cannot plan perfectly. One way of addressing this is to assume that a specific adaptation decision is made (perhaps after a lag), and then estimate costs and residual impacts under a range of possible future climates. By assigning likelihoods to these possible climate outcomes,[2] it is then possible to estimate the expected costs and impacts under the defined adaptation action. By repeating the analysis with alternative adaptation decisions it is possible to identify a range of possible costs and residual impacts (and also identify the action which copes 'best', using some defined criteria, with climate change uncertainty). However, while it is possible to construct an experimental design which enables the estimation of the costs of adaptation and residual impacts at the scheme or management-unit level (assuming varying degrees of perfect knowledge), it is rather more complicated to estimate costs across a region or indeed the globe.

Such costs, aggregated across a large global domain, could in principle be estimated by: (i) extrapolating from a small number of case studies (bottom-up), or (ii) taking a (top-down) large-scale perspective and estimating at the same time costs and impacts across multiple locations. Unfortunately, adaptation is very locally specific. In most cases there are multiple adaptation options, which vary with local geographical, financial, institutional and socio-economic circumstances. Extrapolation from a small number of case studies could therefore be very misleading (and the studies would be likely to use different climate scenarios), and it is impossible in practice to take a realistic top-down approach. Extrapolation has been used, however, to construct climate response functions relating impact (and implicitly adaptation costs) to climate forcing (e.g. Mendelsohn et al., 2000).

Estimation of the cost of adaptation and residual damages across a large geographic domain therefore needs to adopt a different approach from the estimation of costs and impacts at the system scale. One approach could be to base estimates on current and projected expenditure on management infrastructure, assuming that an additional fraction represents the additional cost of climate change. However, such an approach under-estimates the cost of adaptation where expenditure is currently low (where the adaptation deficit is highest), and does not account for the residual damages.

An alternative is therefore to take a top-down approach using a suite of generic indicative adaptation options, and estimating changes in hydrological characteristics using some form of macro-scale hydrological model. For example, the costs of adaptation to changes in water supply availability due to climate change could be indexed by the costs of providing additional storage capacity to maintain supply reliability. The costs of adapting to altered flood risk could be characterised by estimating the costs of providing flood protection to a target standard of service. In both cases, costs would be estimated by applying generalised cost functions (dollars per megalitre of storage, for example, or dollars per metre of flood protection embankment). Such an approach would give an indication of the potential magnitude of adaptation and residual damage costs in a consistent way, although would not be precisely accurate.

3.4 Case studies at the system scale

Very few estimates of the cost of adaptation to climate change in the water sector at the system level have been published. Many estimates are likely to have been made – as they have been for water supply in England and Wales, for example – but in most cases reports are confidential for business reasons.

Among the few published examples, Zhu et al. (2007) used a dynamic programming method to estimate the optimum floodplain adaptation strategy in Sacramento, California, under different

3

climate and urbanisation assumptions. With no urbanisation, they estimated the present value (over 150 years) of flood control costs and residual damages to be $392 million in the absence of climate change, and $485 million under a climate change scenario, giving an additional cost (present value) of $93 million due to climate change. With a high urbanisation rate, the corresponding costs are $828 million without and $1031 million with climate change, giving a cost of $203 million due to climate change.[3] The difference between the two sets of estimates emphasises the role of additional drivers in influencing the costs of adaptation. Zhu et al. (2007) noted that their study assumed perfect foreknowledge of climate change trends.

Medellin-Azuara et al. (2008) used a similar dynamic programming method to identify optimum water supply management measures under current and future climatic conditions (as characterised by one climate model), assuming perfect foreknowledge. They found that the scenario considered did not require any major capital investments, but led to increases by 2050 in annual operating costs of $369 million per year (essentially the cost of adapting pumping and treatment regimes to altered volumes and timing of flows). They also calculated that water scarcity costs – primarily in terms of agricultural impacts – would increase by $121 million per year; these are the residual impacts left over after adaptation. In this example, the residual impacts are therefore approximately a third of the adaptation costs, although it is not self-evident how general this ratio is.

These studies did not attempt to extrapolate from the system that was investigated. A study of the potential costs of adaptation to altered water quality and stormwater flood risks in England and Wales (ICF, 2007) sought to estimate national costs using a small number of case studies (bottom-up generalisation, as outlined in Section 3.3 above). This concluded that the costs of adapting to altered stormwater flood risks – as characterised by increasing storage requirements – ranged between £0.9 billion and £1.1 billion per year (at discount rates of 5–6% and over 40 years), equivalent to approximately 25% of current stormwater drainage expenditure. On the other hand, the costs of increasing effluent quality to maintain water quality standards was only between £4 million and £25 million per year, depending on standards and change in river flows.

Kirshen et al. (2005) described a methodology to estimate the costs of changes in water supply in China under climate change, using a variation on the top-down methodology introduced in the previous section. Runoff was simulated in each major basin in China under current and two future climates, and the costs of additional storage and groundwater development necessary to maintain target yields were estimated using generalised cost functions relating dollars to units of storage. The study assumed perfect foreknowledge. Although summary cost data were not presented, in the Huang Ho River it was calculated that annualised costs (50-year time horizon and 3% discount rate) of meeting present demand would increase from approximately $200 million by the 2050s with no climate change, to approximately $700 million under one climate change scenario (a climate change effect of approximately $500 million); under another scenario, it proved impossible to meet present demand.

3

3.5 Estimates of the costs of adaptation at the global scale

There have been two global-scale assessments of the potential cost of climate change in the water sector, one examining water supply to domestic, industrial and agricultural consumers (Kirshen, 2007, and as subsequently revised to provide the basis for the estimate in the UNFCCC report (UNFCCC, 2007)), and one focusing on the altered costs of providing irrigation water (Fischer et al., 2007).

The Kirshen/UNFCCC study assumed a generic set of adaptation responses to water shortage, and used generalised cost functions to estimate the costs of these adaptation responses. Estimates of the need for adaptation were made by changing national water resource availability in accord with large-scale scenarios for change in national rainfall, derived from a set of scenarios based on climate models used in the IPCC AR4 (they did not use hydrological models to estimate change in generic reservoir performance). The study compared resource availability with demand for water from domestic, municipal, industrial and agricultural consumers, plus aquatic ecosystems. Resource availability included both surface and groundwater resources, and the study used a simple prioritisation procedure to determine, for each country, a feasible combination of supply-side measures (new reservoirs or groundwater abstraction, and desalination) and demand-side measures (increased water-use efficiency). If it was not possible to meet demands from all possible supplies, then it was assumed irrigation demands were left unmet – a measure of residual damage. The study concluded that the cost of adaptation to both population and climate change in the water sector, across the globe, totalled $531 billion to 2030 (under an A1b scenario), including both economic and climate change. This figure was increased to $898 billion in the UNFCCC report to account for use of more expensive sites and unmet irrigation demands. Approximately a quarter of this is assumed to be specifically for climate change (UNFCCC, 2007) – based on a case study in West Africa which explicitly calculated costs with and without climate change – giving a cost of approximately $225 billion to 2030, or $11 billion per year.

This study represents a coherent attempt to estimate costs of increasing water supply capacity, using best-available (at the time) estimates of the cost of generic adaptation measures; it takes the top-down approach outlined in the previous section. However, the total is likely to be a considerable under-estimate, for several reasons (most acknowledged in Kirshen, 2007). The study worked at the national level, and assumed that water resources would be transferred within a country from areas of surplus to areas of deficit. For small countries, this is not problematic; for large and diverse countries, however, this is unrealistic. For example, although water resource availability at the national scale in China may not vary much with climate change, large areas of China are likely to suffer reduced availability. As noted above, Kirshen et al. (2005) estimated that annual costs of adapting to climate change in just the Huang Ho basin could be $0.5 billion by 2050 – in other words, costs in just one river basin account for approximately 5% of the estimated *global* costs. The study used an empirical relationship between annual runoff and its variability, and reservoir storage capacity (McMahon et al., 2007). The relationship cannot incorporate the effects of changes in the timing of runoff through the year due to climate change, which would also be expected to alter the storage capacity required to maintain a target yield and reliability.

3

The use of an average climate-change scenario, rather than individual scenarios, may give a biased – probably downwards – estimate of the costs of climate change. The mean response from several climate scenarios is likely to be higher than the response from the mean of the scenarios, because the loss function is asymmetrical: a reduction in availability is worth more than an increase in availability. The study also did not consider the residual damages: it assumed perfect adaptation to maintain a notional standard of service, so under the study assumptions arguably these residual damages could be assumed to remain approximately constant. However, because adaptation will be imperfect, residual damages would not be zero in practice.

Changes in operating costs may also be substantial (in the California example above, annual operating costs increased by $0.4 billion/year due to climate change), and are not included because the focus of the study was on investment needs. In the opposite direction, the study focused on the costs of augmenting supplies; demand-side measures to maintain a reliable supply–demand balance are typically cheaper than supply-side measures. Finally, the Kirshen study provided an estimate of the public investment costs (by water suppliers, not users) to meet socio-economic and climate changes by 2050. The UNFCCC report assumed that 25% of these costs represents the fraction attributable to climate change alone; the basis for this assumption is not clear.

It is not possible to assess the extent of potential under-estimation of the costs of maintaining water supplies without undertaking a more detailed analysis taking into account multiple scenarios and working at the finer spatial scale. It is also important to emphasise that the adaptation-cost figure characterises only the cost of adapting to water shortage. It does not include costs of flood management, storm drainage, water quality enhancement, hydropower generation, navigation or maintaining ecosystem services.

The irrigation study (Fischer et al., 2007) simulated future irrigation demands without climate change and under two climate models and two emissions scenarios (representing 'no policy' (SRES A2) and 'mitigation' (SRES B1)), using the FAO agro-ecological zones model. Again, they used a top-down methodology similar in principle to that outlined above. By 2080, unmitigated climate change would increase the cost of providing additional irrigation – in terms of capital infrastructure and operating costs – by $24–27 billion per year; the mitigation scenario reduced these costs by $8–10 billion per year. The 'unmitigated' adaptation costs are considerably higher than the approximately $11 billion calculated by Kirshen/UNFCCC for water supply as a whole, but the Fischer et al. figure relates to 2080 rather than spend to 2030.

There is currently no global-scale information on the relative costs of adaptation among the different components of water resources management. However, it is highly likely that costs of storm and river flood management will be very significant (globally), even if costs of water quality management, navigation enhancement and ecosystem protection are not large relative to the costs of maintaining water supplies.

3.6 Conclusions

The UNFCCC estimates apply a broadly robust methodology to estimate the potential costs of meeting the challenges to water supplies posed by climate change. However, the application of the methodology – given the time available for the study – necessitated a few approximations (outlined above) which are likely to mean that the headline global figure of $9–11 billion/year *additional* public investment requirements to cope with climate change is an under-estimate of the costs of adaptation. The residual impacts not covered by adaptation are likely to be high, although again unquantified, largely because adaptation will be imperfect and lagged. The figure also represents the cost of adapting to climate change, assuming no adaptation deficit.

1 Annex I countries are the 36 industrialised countries and Economies in Transition that have accepted emission 'caps' under the UNFCCC that limit their total greenhouse-gas emissions within a designated timeframe.

2 This is non-trivial, and potentially problematic. There is the practical problem that it can be difficult to define credibly the probability distribution of future impacts, because this distribution may be very much influenced by assumptions about distributions of the driving causes of uncertainty. Conceptual problems arise because it is not possible to assign likelihoods to some aspects of uncertainty (such as future emissions).

3 They do not publish the breakdown between adaptation costs and residual damages.

References

Agrawala, S., Crick, F., Jette-Nantal, S. and Tepes, A. (2008) 'Empirical estimates of adaptation costs and benefits: a critical assessment'. In Agrawala, S. and Fankhauser, S. (eds) *Economic Aspects of Adaptation to Climate Change*. OECD.

Briscoe, J. (1999) 'The financing of hydropower, irrigation and water supply infrastructure in developing countries', *International Journal of Water Resources Development* 15 (4): 459–491.

Fischer, G. et al. (2007) 'Climate change impacts on irrigation water requirements: effects of mitigation, 1990–2080', *Technological Forecasting and Social Change* 74(7): 1083–1107.

ICF (2007) *The Potential Costs of Climate Change Adaptation for the Water Industry*. Report to Environment Agency.

Kirshen, P. (2007) *Adaptation Options and Cost in Water Supply*. Report to UNFCCC Secretariat Financial and Technical Support Division.

Kirshen, P., McCluskey, M., Vogel, R. and Strzepek, K. (2005) 'Global analysis of changes in water supply yields and costs under climate change: a case study in China', *Climatic Change* 68(3): 303–330.

McMahon, T.A., Pegram, G.G.S., Vogel, R.M. and Peel, M.C. (2007) 'Revisiting reservoir storage-yield relationships using a global streamflow database', *Advances in Water Resources* 30: 1858–1872.

Medellin-Azuara, J. et al. (2008) 'Adaptability and adaptations of California's water supply system to dry climate warming', *Climatic Change* 87: S75–S90.

Mendelsohn, R., Morrison, W., Schlesinger, M.E. and Andronova, N.G. (2000) 'Country-specific market impacts of climate change', *Climatic Change* 45: 553–569.

UNFCCC (2007) *Investment and Financial Flows to Address Climate Change*. Climate Change Secretariat, Bonn.

Zhu, T. et al. (2007) 'Climate change, urbanisation and optimal long-term floodplain protection', *Water Resources Research* 43, W06421, doi: 10.1029/2004WR003516.

4

Adaptation costs for human health

Sari Kovats

4

Summary

- All countries will need to implement measures to reduce or avoid the additional impacts on health due to global climate change.

- The magnitudes of costs of adaptation in the health sector are unknown but potentially large. Not implementing additional adaptation will be even more expensive in terms of the burden of additional injuries, illnesses and deaths on society.

- The UNFCCC has estimated costs of $4–12 billion for adaptation in the health sector in developing countries. These costs represent the costs of preventing additional cases of malnutrition, malaria and diarrhoeal disease due to climate change by 2030.

- The UNFCCC costs do not consider the full range of climate futures or the full range of disease/health outcomes that will be affected by climate change. On balance, the UNFCCC costs are likely to be an under-estimate of the full costs of adaptation in the health sector in developing countries.

- High-income countries will also bear costs to adapt to climate change, including the high costs associated with adaptation of health systems infrastructure.

- The current lack of investment in public health is a considerable barrier to adaptation. However, investment alone is not sufficient to improve health in developing countries, as many other barriers remain, such as poor governance, inequality and low adaptive capacity.

4.1 Introduction and scope

The potential impacts of climate change on population health include a wide range of diseases and health outcomes, from infectious diseases to malnutrition and disaster-related injuries (Confalonieri et al., 2007). Adaptation, broadly defined, would include all activities that reduce or prevent these 'additional' cases or deaths. Several reviews of such adaptation strategies, policies and measures have now been published (Menne and Ebi, 2005; Ebi, Kovats and Menne, 2006). Further, initiatives are now underway to support adaptation in the health sector in low- and middle-income countries.

This chapter provides an overview of the current literature on adaptation costs for the 'health sector'. The health sector is here limited to conventional public health activities, although it is well recognised that adaptation in other sectors is probably more important for reducing the health impacts of climate change (through disaster mitigation, food and water security, and providing decent infrastructure). At the time of writing (July 2009), there is only one set of comprehensive adaptation costs for health and these are provided in the UNFCCC report and related journal paper (UNFCCC, 2007; Ebi, 2008).

Health 'adaptation costs' include:

- costs of improving or modifying health protection systems to address climate change, for example, expanding health or vector surveillance systems – this includes the costs associated with building new infrastructure, training new health care workers, and increasing laboratory and other capacities
- costs of introducing novel health interventions (e.g. heat-wave warning systems)
- additional costs for meeting environmental and health regulatory standards (e.g. air quality standards, water quality standards)
- costs of improving or modifying health systems infrastructure, for example adapting hospitals to hotter summers
- occupational health costs, for example measures to prevent the adverse impacts of increased heat load on the health and productivity of workers
- costs of health research on reducing the impact of climate change, for example evaluation studies
- costs of preventing the additional cases of disease due to climate change as estimated by scenario-driven impact models.

This last type could also be considered as 'damage' or impact costs – rather than adaptation costs (OECD, 2008). The health costs presented in the UNFCCC report are of this type (UNFCCC, 2007). Given that many of the relevant preventive activities are outside the remit of the health sector, and are currently not implemented, a key role for the health sector will be to simply prevent or treat the additional disease cases caused by climate change.

4.2 The UNFCCC estimates for health adaptation costs

The UNFCCC report estimated the adaptation costs for the health sector to be in the range of $4–12 billion per year in 2030 (UNFCCC, 2007). The adaptation costs are for preventing the additional climate-change-related cases of diarrhoeal disease, malnutrition and malaria in 2030. (These are discussed in more detail below.) The total costs were estimated by taking the number of additional cases and multiplying them by the costs of prevention per child, obtained from the Disease Control Priorities in Developing Countries project (World Bank, 2006). Thus, uncertainties relate to both the estimates of the future burden and the method for costing.

The prevention activities were based on currently deployed interventions and are listed in Table 4.1. The costs did not include the cost of implementing programmes, including health care personnel costs or infrastructure costs. The costs of initiating programmes in new areas can be significant. Further, such costs would occur with a shift in disease distribution, even if there was no net increase in the number of cases.

Table 4.1 Health 'intervention sets' included in the UNFCCC report

Outcome	Intervention sets	Average cost per child ($), 2001
Diarrhoeal disease	Breastfeeding promotion, rotavirus and cholera immunisation	15.09
	Improvement in water supply and sanitation	53.00
Malaria	Insecticide-treated bed-nets plus case management with artemesin-based combination therapy	88.50
	Indoor residual insecticide spraying plus above	123.05
Malnutrition	Breastfeeding promotion, child survival programmes, nutritional programmes, growth monitoring	17.40 to 23.09

As Table 4.1 shows, the costs of interventions for child health, while requiring large investments to meet the MDGs, result in costs per child that are relatively low. Water and sanitation interventions are highly cost-effective (Hutton, Haller and Bartram, 2007). Further, improvements in water supply and sanitation, and other environmental interventions, have multiple benefits (e.g. for education and welfare), not all of which are accounted for.

The costs of the nutrition interventions are likely to be under-estimated as they address only a narrow range of public health measures. The annual per capita cost of providing food to improve child health in Africa has been estimated to be much higher (Edejer et al., 2005). Ebi (2008) estimated that including such costs would increase the adaptation costs more than 10-fold. The conservative approach used also avoids any double counting with the agriculture sector adaptation costs that include the cost of feeding more people (see Chapter 2, Wheeler and Tiffin, in this report).

Table 4.2 Projected costs in 2030 to manage climate change-related cases of diarrhoeal diseases, malnutrition, and malaria for three climate scenarios (million US$)

Emissions scenario	Diarrhoeal diseases Middle	Diarrhoeal diseases High	Malnutrition Middle	Malnutrition High	Malaria Middle	Malaria High
Stabilisation of emissions at 550ppm CO_2 equivalent	1706	6024	53.9 – 71.5	112.9 – 149.9	1573 – 2145	3236 – 4515
Stabilisation of emissions at 750ppm CO_2 equivalent	1983	6814	81.3 – 107.9	162.5 – 215.6	1928 – 2691	3994 – 5573
Unmitigated emissions	2731	9010	62.2 – 82.6	125.2 – 166.2	3059 – 4269	6293 – 8781

Note: the three climate scenarios are relative to baseline climate. High, middle and low costs (low estimates not shown) reflect uncertainty around the relative risk of increased cases due to climate change

Source: Ebi (2008)

The analysis makes a number of necessary but unlikely assumptions. The number of annual cases of diarrhoeal disease, malaria and malnutrition is assumed to remain the same over time. Population growth is projected to increase under the medium variant from 6.1 billion in 2000 to 8.3 billion in 2030. Using the current number of cases in the analysis assumes that incidence will decrease as population increases, without attribution of the possible reasons for such a decline. If disease incidence rates remain constant, or decline less rapidly, until 2030, then the number of cases attributed to climate change would increase in these estimates (and therefore the costs would increase).

The estimates also assume that the cost of the interventions remain constant from 2000 to 2030. There are further uncertainties about the unit cost, which varies significantly between countries (World Bank, 2006). It should also be noted that new interventions may be developed in the next few years (e.g. a vaccine for malaria) or current ones may become ineffective (e.g. through widespread insecticide resistance in mosquitoes). The health interventions in Table 4.1 are not 100% effective (based on research on both efficacy and compliance) (Edejer et al, 2005). Therefore, they will not prevent all the additional cases attributed to climate change, and there will be some 'residual' health impacts.

4

4.3 The burden of disease due to climate change

The estimated additional cases are derived from the WHO Global Burden of Disease (GBD) project (Lopez et al., 2006; McMichael et al., 2004). The limited number of health outcomes in this assessment was due to the lack of exposure-response relationships and other information needed to estimate future health burdens due to climate change.

The estimates of future health impacts are dependent upon the future worlds (development pathways), which makes presentation of costs difficult as they have to be contextualised within specific sets of scenarios and assumptions. The GBD estimates were derived from three emissions scenarios (Table 4.2), that included two stabilisation scenarios (Arnell et al., 2002) and one business-as-usual scenario that corresponds to the IPCC IS92a emissions scenario.

The estimates for additional cases of diarrhoeal disease were based on the published evidence for the direct effect of temperature on hospital admissions for diarrhoeal disease. No estimates were available for the impact of changes in rainfall patterns on diarrhoeal disease because of the lack of published evidence – and the complexity of the system. Similarly, the estimates for malnutrition cases are based on large-scale crop modelling that does not include extreme events such as severe drought (Parry et al., 1999). The malnutrition cases make the biggest contribution to the climate-change-attributable burden of disease, but are also the most uncertain.

The estimates for additional cases incorporate some assumptions regarding adaptation (Campbell-Lendrum and Woodruff, 2006; McMichael et al., 2004). For example, the estimates for malnutrition incorporate the assumptions for technological adaptation in the agricultural sector, and the estimates for diarrhoeal diseases assume that the disease is reduced as countries get richer. The original authors of the WHO study estimated that the three outcomes represent approximately 30–50% of the total burden attributable to climate change in low- and middle-income countries in 2030 (McMichael, Bertollini, 2009, personal communication).

4.4 The 'adaptation deficit'

The current burden of climate-sensitive disease is high (Ezzati et al., 2005; Mills and Shillcutt, 2004). The 'adaptation deficit' is therefore a key issue in health impacts as the greatest burden of climate-sensitive diseases is in low-income countries. The current 'level of health service provision' is inadequate in many countries. The target levels of provision are either seen as what is feasible (e.g. Millennium Development Goals) or what is ideal (e.g. Health for All). As well as the moral arguments for improving health, there are also economic ones. The WHO established its Commission on Macroeconomics and Health to provide the evidence that poor health impedes economic development (Sachs, 2001). Despite recent improvements in bilateral aid, the development of targeted programmes (e.g. Roll Back Malaria) and the new philanthropic ventures (the Global Fund, Bill and Melinda Gates Foundation), many countries still lack the investment needed to achieve the health-related Millennium Development Goals (Mills and Shillcutt, 2004).

4

As discussed in Chapter 1 of this report (Fankhauser), there is often no clear distinction between strategies for development and strategies for adaptation. Many current public health and disease control measures, if implemented successfully, will reduce the impact of climate change on human health. This message has recently been reinforced by the World Health Organization which recommended that countries should strengthen their health systems as well as integrating health measures into plans for adaptation to climate change (WHO Executive Board, 2008).

4.5 Costing health: methods

The 'burden of disease' is the term used to describe the total impact of disease or health condition in a population, including deaths, cases and years lived with disability (for chronic diseases). The metrics used in environment and health decision making include: deaths, DALYS (disability-adjusted life years), QALYs (quality-adjusted life years) (Hofstetter and Hammitt, 2001; Mathers et al., 2003). Methods of cost-benefit analysis (CBA) and cost-effectiveness analysis (CEA) are not generally applied to health issues at the global scale but are used to assess the benefits of alternative policy choices at the local scale. The main criticism of the costing method in the UNFCCC report is that micro-methods (which are very context-specific) are being applied inappropriately to a global problem.

Health has been incorporated into some integrated assessment models (IAMs) as a damage function or an impairment to economic productivity (Bosello, Roson and Tol, 2006; Tol, 2002). Health costs, when estimated, make a considerable contribution to overall damage costs, as the statistical value of a life is high. In general, the IAMs have incorporated into their damage estimates only the very basic health models for heat/cold effects on mortality and malaria.

There have been several reviews of the costs of health interventions, where cost-effectiveness is the main method used. The Disease Control Priorities in Developing Countries project has evaluated the scientific and economic evidence on the available interventions for all the major infectious and chronic (non-communicable) diseases (World Bank, 2006). Thus, there have been extensive reviews of the cost-effectiveness of health interventions relevant to climate change in the following areas:

- malaria
- malnutrition
- diarrhoeal disease
- food safety.

However, there is still considerable uncertainty around the cost-effectiveness of these interventions. Some have been relatively well researched, e.g. insecticide-treated bed-nets, but for some there is less information (e.g. indoor residual insecticide spraying) (Mills and Shillcutt, 2004). For some interventions, such as heat-health warning systems, there is practically no robust information on their effectiveness (Kovats and Hajat, 2008). It should also be noted that health adaptation costs in high-income countries are not included in the UNFCCC costs. Such costs would be higher due to the higher infrastructure and labour costs. In addition, the thresholds for intervention (costs per case prevented) are higher.

4

4.6 Case studies

There have been increasing numbers of studies that quantify climate change impacts at the national or local level, but very few estimate the costs of these health effects, either as direct health costs (deaths, welfare costs) or in terms of health service usage. However, in most assessments some consideration of adaptation is included. For example, the PESETA study monetised the impact of climate change on heat- and cold-related mortality and on food poisoning in Europe, assuming some acclimatisation (Watkiss et al., 2007). At the time of writing (July 2009) there are no comprehensive 'global' studies of the health impacts or adaptation costs.

No case studies of health adaptation costs have been published in high-income countries. However, many countries, including the UK, have implemented health adaptation measures in the form of heat-wave plans (Department of Health, 2008). The costs of these plans can be estimated. Heat early-warning systems are relatively inexpensive, unless they include active measures that are implemented following an alert. The reported costs for European countries range from €200,000 to €6 million per year (WHO Regional Office for Europe, 2008). Structural interventions are more expensive. France spent more than €150 million in 2004 on providing additional staff and cool rooms in residential homes for the elderly (Michelon, Magne and Simon-Delaville, 2005). Mortality in this group doubled during the 2003 heat wave. Estimates for the costs of adapting dwellings in London to hotter summers are discussed in Section 6.5 in this report.

In low- and middle-income countries, a recent review for the World Health Organization found very few examples of studies estimating the costs of adaptation (Markandya and Chiabai, 2008). An unpublished study in South Africa quantified adaptation costs as the prevention costs of the additional burden of malaria cases due to climate change to 2025 (van Rensburg and Blignaut, 2002). However, several such case studies are currently being undertaken in African and Asian countries.

Health projects are now being included in the National Adaptation Programmes of Action (NAPAs) but they typically address current disease control issues, usually for malaria. Financial requests for specific projects range from $22,000 to $600,000 for disease-control, early-warning information systems to $7 million for a major project in Bangladesh. Environmental projects such as rainwater harvesting and improving food security are not included as health projects although they will have health benefits. Oxfam selected some of the NAPA projects (including one form Samoa on developing climate-based health early-warning systems) in order to scale up the adaptation costs (Oxfam, 2007). As discussed in Chapter 1 of this report (Fankhauser), the NAPA projects capture only a very small part of adaptation needs and adaptation costs.

4.7 Conclusions

In conclusion, the UNFCCC costs are an under-estimation of the total 'health sector' costs because of all the activities, diseases and countries that are not included in these estimates. The UNFCCC costs are for impacts that consider incremental improvements in health and economic development, but not explicit measures that might be implemented in response to the threat of climate change.

The methods used to generate the UNFCCC health costs were based on the best available information at the time for developing countries. This evidence base is still very limited and it is difficult to see how a more comprehensive study could have been undertaken in the short timeframe available. There are many uncertainties in the UNFCCC health costs, and the most important of these are summarised in Table 4.3.

Table 4.3 Summary of uncertainties in the UNFCCC adaptation costs of health in developing countries in 2030

Issue	Comments	Likely effect on total final costs
Burden of disease limited to 3 health outcomes – diarrhoeal disease, malaria, malnutrition	Climate change is likely to affect other infectious diseases but the magnitude is uncertain	Under-estimate
Number of cases assumes a decline in incidence	Highly uncertain as incidence is declining in some countries, but many will fail to meet their MDG targets.	Unknown
Choice of intervention sets	Limited range of options, could include a wider range of measures	Under-estimate
Costs of interventions, and the assumption of no changes to cost over time	Costs vary between countries and are likely to change over time.	Under or over-estimate

Acknowledgements

I would like to thank the following for helpful review comments:

- Diarmid Campbell-Lendrum, World Health Organization, Geneva
- Guy Hutton, World Bank
- Kristie Ebi, IPCC Working Group II Technical Support Unit
- Anil Markandya, Aline Chiabai, Elena Ojea, Julia Martin-Ortega, Basque Climate Change Centre
- Tony McMichael, Australian National University.

4

References

Arnell, N.W., Cannell, M.G., Hulme, M., Kovats, R.S., Mitchell, J.F., Nicholls, R.J., Parry, M., Livermore, M.T. and White, A. (2002), 'The consequences of CO_2 stabilization for the impacts of climate change', *Climatic Change*, vol. 53, pp. 413–446.

Bosello, F., Roson, R. and Tol, R.S. 2006, 'Economy-wide estimates of the implications of climate change: human health', *Ecological Economics*, vol. 58, pp. 579–591.

Campbell-Lendrum, D. and Woodruff, R.E. (2006), 'Comparative risk assessment of the burden of disease from climate change', *Environmental Health Perspectives*, vol. 114, no. 12, pp. 935–1941.

Confalonieri, U., Menne, B., Akhtar, R., Ebi, K.L., Hauengue, M., Kovats, R.S., Revich, B., and Woodward, A. (2007) 'Human health' in M.L. Parry et al. (eds) *Climate Change 2007: Impacts, Adaptation and Vulnerability. Contribution of Working Group II to the Fourth Assessment Report of the Intergovernmental Panel on Climate Change*. Cambridge University Press, Cambridge, pp. 391–431.

Department of Health (2008) *Heatwave Plan for England – Protecting health and reducing harm from extreme heat and heatwaves*. Department of Health, London.

Ebi, K.L. (2008) 'Adaptation costs for climate change-related cases of diarrhoeal disease, malnutrition, and malaria in 2030', *Globalization and Health*, vol. 4, no. 9.

Ebi, K.L., Kovats, R.S. and Menne, B. (2006) 'An approach for assessing human health vulnerability and public health interventions to adapt to climate change', *Environmental Health Perspectives*, vol. 114, no. 12, pp. 1930–1934.

Edejer, T.T.T., Alkins, M., Black, R., Wolfson, L., Hutubessy, R. and Evans, D.B. (2005) 'Cost effectiveness analysis of strategies for child health in developing countries', *British Medical Journal*, vol. 331.

Ezzati, M., Utzinger, J., Cairncross, S., Cohen, A.J. and Singer, B.H. (2005), 'Environmental risks in the developing world: exposure indicators for evaluating interventions, programmes, and policies', *Journal of Epidemiology and Community Health*, vol. 59, pp. 15–22.

Hofstetter, P. and Hammitt, J.K. (2001) *Human Health Metrics for Environmental Decision US EPA Office of Research and Development Support Tools: Lessons from Health Economics and Decision Analysis. EPA/600/R-01/104*, Environmental Protection Agency, Washington DC.

Hutton, G., Haller, L., and Bartram, J. (2007) 'Global cost-benefit analysis of water supply and sanitation interventions', *Journal of Water and Health*, vol. 5, no. 4, pp. 481–502.

Kovats, R.S. and Hajat, S. (2008) 'Heat stress and public health: a critical review', *Annual Review of Public Health*, vol. 29, pp. 41–55.

Lopez, A.D., Mathers, C.D., Ezzati, M., Jamison, D.T. and Murray, C.J. (2006) 'Global and regional burden of disease and risk factors, 2001: systematic analysis of population health data', *Lancet*, vol. 367, pp. 1747–1757.

Markandya, A. and Chiabai, A. (2008) *Assessment of the Likely Financial Costs Necessary for Health Protection for Climate Change. Technical Report for WHO Global consultation on 'Guiding Research to improve health protection from climate change'*. World Health Organization, Geneva.

Mathers, C.D., Murray, C.J., Ezzati, M., Gakidou, E., Salomon, J.A. and Stein, C. (2003), 'Population health metrics: crucial inputs to the development of evidence for health policy' (editorial), *BioMed Central*.

McMichael, A.J., Campbell-Lendrum, D., Kovats, R.S., Edwards, S., Wilkinson, P., Edmonds, N., Nicholls, N., Hales, S., Tanser, F.C., Le Sueur, D., Schlesinger, M. and Andronova, N. (2004), 'Climate change' in M. Ezzati et al. (eds) *Comparative Quantification of Health Risks: Global and Regional Burden of Disease due to Selected Major Risk Factors. Vol.2*. World Health Organization, Geneva, pp. 1543–1649.

Menne, B. and Ebi, K.L. (2005) *Climate Change and Adaptation Strategies for Human Health*. Steinkopff Verlag, Darmstadt, Heidelburg.

Michelon, T., Magne, P. and Simon-Delaville, F. (2005) 'Lessons of the 2003 heat wave in France and action taken to limit the effects of future heat waves,' in W. Kirch, B. Menne and R. Bertollini (eds) *Extreme Weather Events and Public Health Responses*. Springer-Verlag, pp. 131–140.

Mills, A. and Shillcutt, S. (2004) 'Communicable diseases' in B. Lomborg (ed.) *Global Crisis, Global Solutions*. Cambridge University Press, Cambridge, pp. 62–114.

OECD (2008) *Economic Aspects of Adaptation to Climate Change: Costs, Benefits and Policy Instruments*. OECD, New York.

Oxfam (2007) *Adapting to Climate Change: What's Needed in Poor Countries, and Who should Pay*. Oxfam, Oxford.

Parry, M.L., Rosenzweig, C., Iglesias, A., Fischer, G. and Livermore, M.T. (1999), 'Climate change and world food security: a new assessment', *Global Environmental Change*, vol. 9, pp. S51–S67.

van Rensburg, J.J.J. and Blignaut, J.N. (2002) 'The economic impact of an increasing health risk due to global warming', Forum for Economics and the Environment – First Conference Proceedings. Forum for Economics and the Environment, Cape Town, pp. 17–127

Sachs, J. (2001), *Macroeconomics and Health: Investing in Health for Economic Development. Report of the Commission on MacroEconomics and Health*. World Health Organization, Geneva.

Tol, R.S. (2002), 'Estimates of the damage costs of climate change. Part I. Benchmark estimates', *Environmental and Resource Economics*, vol. 21, pp. 47–73.

UNFCCC (2007), *Background Paper on Analysis of Existing and Planned Investment and Financial Flows Relevant to the Development of Effective and Appropriate International Response to Climate Change*. UNFCCC, Bonn.

Watkiss, P., Horrocks, L., Pye, S., Searl, A. and Hunt, A. (2007) *Projection of Economic Impacts of Climate Change in Sectors of Europe based on Bottom up Analysis (PESETA). Human Health. Summary for Policy Makers: Report to JRC/IPTS*.

WHO Executive Board (2008) *Climate Change and Health: Report by the Secretariat*. WHO, Geneva, EB122/4.

WHO Regional Office for Europe (2008), *Improving Public Health Responses to Extreme Weather/Heat-waves - EuroHEAT. Technical Summary*. WHO Regional Office for Europe, Copenhagen.

World Bank (2006) *Disease Control Priorities in Developing Countries* (2nd edn). The World Bank/Oxford University Press, Washington DC.

5

Adaptation costs for coasts and low-lying settlements

Robert Nicholls

5

Summary

- The UNFCCC study (Nicholls, 2007) is one of several that have attempted to estimate the costs of adaptation due to sea-level rise, starting with IPCC CZMS (1990).

- All of these studies draw on the extensive experience of 'hold the line' with traditional coastal engineering responses (dykes and beach nourishment). Given the long experience of coastal engineering measures, these costs are useful guides to the required investment, as long as the residual damage is considered (not everywhere will be protected).

- Other aspects of climate change were not considered, such as protection against more intense tropical storms and hurricanes. Response costs would rise significantly under this scenario. Aspects of disaster preparedness and institutional capacity to respond to these extreme events will also need to be considered.

- For coastal ecosystems, dyke construction and upgrade will enhance losses via coastal squeeze, and sustaining the environmental functions and human safety in coastal zones represents a significant challenge to coastal management which is not included in the UNFCCC cost estimates.

- The UNFCCC estimate of additional costs of $4–11 billion/year is a reasonable estimate of the coastal engineering protection measures that would be required, assuming a 50-year planning horizon and that there is no adaptation deficit.

- However, the costs are under-estimates if we consider responses to high-end sea-level rise scenarios, and other climate change such as more intense storms – as an indicative estimate, costs three times greater than those reported by the UNFCCC might result.

- Residual damage is included in the UNFCCC report in terms of sea flood and land loss estimates of $1–2 billion/year. However, this is incomplete as sea-level rise produces other damages, including significant environmental damages. Countering these additional damages is feasible, but will further raise adaptation costs.

5.1 Introduction

Sea-level rise is one of the issues that brought human-induced climate change to the fore due to the large concentration of settlements and economic activity in low-lying coastal areas. The issue has been extensively assessed since the 1980s with the spectre of millions of environmental refugees as a worst-case impact. Adaptation needs and costs were considered from the beginning, drawing on the extensive experience of coastal engineering and management, including on subsiding coasts. The global costs of protecting developed coasts against sea-level rise were first estimated by IPCC CZMS (1990). There have been updates of these costs based on several different methodologies as outlined below. However, other dimensions of climate change in coastal areas are less assessed and could substantially raise costs, for example in the case of more intense hurricanes, or coastal ecosystem changes such as coral reef degradation due to rising sea surface temperatures and falling pH of ocean waters (Nicholls et al., 2007a).

5

This chapter analyses the various issues, assumptions and methods surrounding existing studies of the global costing of adaptation for coastal areas, including the UNFCCC global assessment of coastal adaptation costs (Nicholls, 2007). The main goal is to consider how these studies might be improved.

5.2 Assumptions and perspectives

What does adaptation in coastal zones involve?

Planned adaptation options to sea-level rise are usually presented as one of three generic options (IPCC CZMS, 1990; Bijlsma et al., 1996; Klein et al., 2001):

1 (Planned) retreat – the impacts of sea-level rise are allowed to occur and human impacts are minimised by pulling back from the coast via land-use planning and development control.

2 Accommodation – the impacts of sea-level rise are allowed to occur and human impacts are minimised by adjusting human use of the coastal zone to the hazard via increasing flood resilience (e.g. raising homes on pilings), warning systems and insurance.

3 Protection – the impacts of sea-level rise are controlled by soft or hard engineering (e.g. nourished beaches and dunes or seawalls), reducing human impacts in the zone that would be impacted without protection. However, a residual risk always remains, and complete protection cannot be achieved. Managing residual risk is a key element of a protection strategy that has often been overlooked in the past.

The three approaches are illustrated in Figure 5.1.

Figure 5.1 An illustration of the possible adaptation responses to sea-level rise

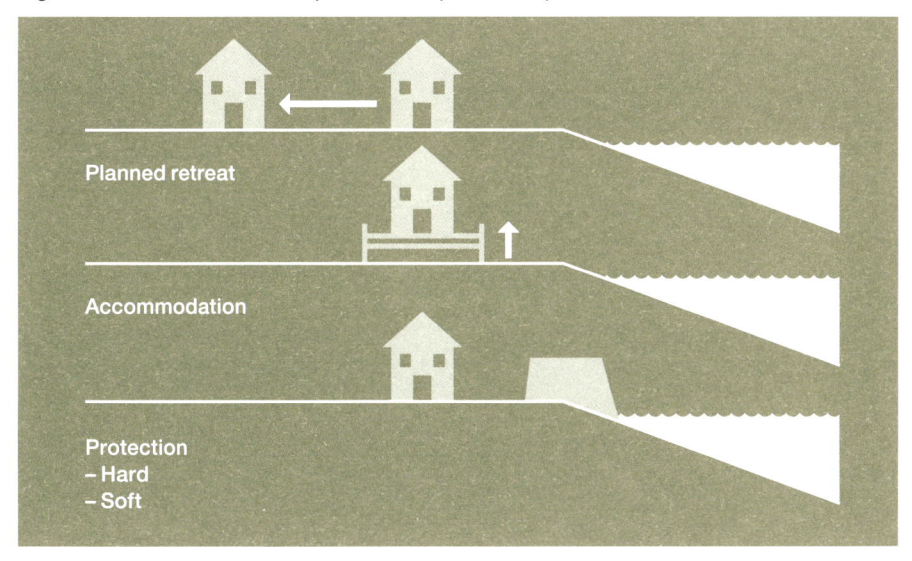

5

Throughout human history, improving technology has increased the range of adaptation options in the face of coastal hazards, and there has been a move from retreat and accommodation to hard protection and active seaward advance via land claim, as illustrated by the changing approaches to managing flooding in the Netherlands (van Koningsveld et al., 2008). Rising global sea level is one factor calling automatic reliance on hard protection into question, and the appropriate mixtures of protection, accommodation and retreat are now being more seriously evaluated.

These methodologies are also appropriate for adapting to more intense storms. For coastal ecosystems, many adaptation measures are in their infancy, and include land-use planning or soft engineering approaches to sustain temperate saltmarshes. For some systems such as coral reefs, no adaptation options are currently available.

What climate scenarios are considered in adaption analyses of coastal zones?

The focus is overwhelmingly on sea-level rise. Actual impacts in coastal zones are a product of *relative* sea-level rise which is the sum of climate-induced changes (global increase in ocean volume and regional effects such as variable thermal expansion, changing mean air pressure and ocean circulation) and non-climate effects on land elevation due to natural processes, such as tectonics and glacial-isostatic adjustment, and, sometimes more importantly, human-induced processes, such as subsidence due to fluid withdrawal, and drainage of coastal soils susceptible to subsidence and oxidation. Hence, relative sea-level rise varies from place to place: it is generally higher than the global mean in areas that are subsiding which includes many populated deltas (e.g. the Mississippi delta), while many coastal cities built in deltaic settings have subsided many metres due to human-induced processes during the 20th century. Examples include Tokyo (up to 5m subsidence), Shanghai (up to 3m subsidence), Bangkok (up to 2m subsidence) and New Orleans (up to 3m subsidence). As well as mean sea-level rise, there is a need to consider changes in extreme sea levels, which will additionally be influenced by more intense storms, if they occur.

Many existing studies of adaptation costs apply global changes of sea level directly with no downscaling to relative sea-level rise. This factor is now being addressed in newer assessment tools such as the DIVA (Dynamic Interactive Vulnerability Assessment) tool (Hinkel and Klein, 2007; Nicholls et al., 2007b; Vafeidis et al., 2008), although the datasets on the non-climate factors require significant improvement. Until recently, no study has addressed changing storms, in part due to the lack of credible scenarios; there is now one paper addressing this (Narita et al., 2009). Similarly, there is little global assessment of coasts and any other climate change factors.

5

What climate impacts are we considering in adaption analyses of coastal zones?

Sea-level rise produces a range of impacts – the major impacts are summarised in Table 5.1. Impact and adaptation analyses normally do not consider all of these impacts, so, by definition, the analyses are incomplete. Historically, analyses have focused on either flooding or land loss and not considered both issues together. Table 5.2 summarises the impacts and adaptation considered in the DIVA model, while Table 5.3 summarises the same information for the FUND coastal module (Tol, 2007).

Table 5.1 The main natural system effects of relative sea-level rise

Natural system effect			Interacting factors	
			Climate	Non-climate
1. Inundation (including flood and storm damage)	a. Surge (from the sea)		Wave/storm climate, erosion, sediment supply	Sediment supply, flood management, erosion, land reclamation
	b. Backwater effect (from rivers)		Run-off	Catchment management and land use
2. Morphological change	a. Wetland loss (and change)		CO_2 fertilisation of biomass production, sediment supply, migration space	Sediment supply, migration space, land reclamation (i.e. direct destruction)
	b. Erosion (of beaches and soft cliffs)		Sediment supply, wave/storm climate	Sediment supply
3. Hydrological change	a. Saltwater intrusion	i. Surface waters	Run-off	Catchment management (over-extraction), land use
		ii. Groundwater	Rainfall	Land use, aquifer use (over-pumping)
	b. Rising water tables/ impeded drainage		Rainfall, run-off	Land use, aquifer use, catchment management

Note: Some interacting factors (e.g. sediment supply) appear twice, as they can be influenced by both climate and non-climate factors

Assessing the costs of adaptation to climate change | 65

5

Table 5.2 The DIVA model: major physical impacts of sea-level rise and the adaptation approaches considered

Physical impacts of sea-level rise		Adaptation approach
1. Inundation, flood and storm damage	a. Surge (sea)	Hard protection (via dykes)
	b. Backwater effect (river)	
2. Wetland loss (and change)		Sediment nourishment
3. Erosion (direct and indirect morphological change)		Beach nourishment
4. Saltwater intrusion	a. Surface waters	'Adaptation not considered'
	b. Groundwater	

Table 5.3 The FUND coastal module – physical impacts of sea-level rise and the adaptation approaches considered

Physical impacts of sea-level rise	Adaptation approach
1. Dry land loss	Hard protection (via dykes)
2. Wetland loss	'Adaptation not considered'

Adaptation analyses at the global scale normally consider one of two distinct approaches:

1 protection of all developed areas (based on some arbitrary definition), such as IPCC CZMS (1990), the Global Vulnerability Assessment (Hoozemans et al., 1993) and the Fast Track Analyses (Nicholls, 2004)

2 an optimisation approach where 'economically worthwhile areas' are defended, such as Fankhauser (1995), Tol (2007) and Sugiyama et al. (2008) – this is normally based on comparing avoided damage and protection costs (DIVA adds beach nourishment to preserve beaches for their touristic value, and nourishment of coastal wetlands to preserve them in situ), and the UNFCCC study used this approach.

Following existing practice, the emphasis in most analyses, including the UNFCCC study, is on preserving the human uses. Protection will tend to degrade coastal ecosystems via coastal squeeze, so this adaptation approach produces secondary impacts which are evaluated in some cases. Hence, protection does not preserve the status quo, but does preserve the valuable dryland.

What is the expected residual damage (and its cost)?

Most studies express residual damages in some form, but not always in economic terms. For many of the optimal analyses, residual damage for the built environment tends to be quite small as most coastal infrastructure and people are concentrated in smaller areas that are more easily protected. They evaluate wetland losses in monetary terms (although this is of course controversial) and it is debatable if this captures the wetland value, especially under a scenario of considerable wetland decline and increasing scarcity. Coastal ecosystem degradation will be significant due to sea-level rise and this will reinforced by protection based on dykes. Nicholls and Klein (2005) identified the twin challenges of maintaining human safety and sustaining coastal ecosystems as a major issue in a European context; this is also true globally.

There is also a residual risk for infrastructure behind defences, and hence in the future we should expect occasional coastal disasters (as seen with New Orleans and Hurricane Katrina in the USA in 2005, or Cyclone Nargis in Myanmar in 2008), even if we adapt optimally. Of course this is a product of climate variability and hence would be true without climate change – the rise in the mean sea level and possibly more intense storms will exacerbate this issue with the result that when floods occur they will be deeper, faster and more likely to cause significant damage to infrastructure and loss of life. Infrastructure losses can be reduced by flood-proofing, while loss of life can be minimised by good warning and evacuation systems.

The UNFCCC study contained estimates of residual land loss per year and residual flood damage (Nicholls, 2007). This was estimated at about $8 billion per year, although without sea-level rise residual damages were estimated at about $6 billion per year. Hence, the increase in residual damage is only about $1–2 billion per year. It should be further noted that subsequent experience with the DIVA tool on other projects suggests that these damage estimates may have been rather conservative, and should be roughly doubled, which raises the additional residual damage due to sea-level rise to $2–3 billion per year (with rounding).

The UNFCCC study did not include residual damages on coastal ecosystems including coastal wetlands such as saltmarshes and mangroves. In many cases adaptation measures are available for such areas, from selective retreat to sediment management and nourishment. This is likely to raise the adaptation costs significantly.

What is the current adaptation deficit?

The published estimates of adaptation costs consider only the incremental cost of climate change (i.e., they assume that there is a good existing infrastructure to upgrade for climate change), which is often not the case, especially in developing countries. The UNFCCC study focused on the incremental costs of climate change, rather than the total cost of adaptation to achieve a significant reduction in vulnerability to all climate factors, including existing variability. This issue has been termed the adaptation deficit (Burton, 2004) and shows that the issues of development and the cost of successful adaptation to climate change are intimately linked.

The question of the current adaptation deficit for coastal areas has not been systematically addressed. DIVA computes an optimum defence standard for dykes based on present conditions. While this has not been systematically compared to actual defences, qualitative assessment of

5

these numbers suggests that in many places, especially the developing world and even in developed-country locations like New York City, there is a substantial gap between the existing and optimum (according to DIVA) defence systems.

There are some cost estimates which may suggest the order of this adaptation deficit. An example is the $50 billion price tag to upgrade New Orleans for hurricanes post-Katrina. Given a 30-year period of upgrade, this translates into $2 billion/year for one coastal city. As the adaptation needs for sea-level rise in 2030 were estimated to be up to only $13 million/year, this single estimate suggests that the adaptation deficit could exceed the incremental costs of sea-level rise. This issue clearly requires further, more comprehensive assessment.

Without the adaptation deficit being made good, in addition to upgrading for climate change, much climate damage will not be avoided. This implies that adaptation from a perspective of coastal management, rather than just climate change, would be much more costly. Based on the few known examples (e.g. New Orleans, New York) the magnitude of the adaptation deficit (assuming a protect response) is likely to exceed the $11 billion/year estimated in the UNFCCC report, assuming that 30 to 50 years are spent on making good this deficit.

5.3 Methods of estimating costs

Most of the available cost estimates are bottom-up estimates based on a long history of coastal management and engineering experience. This draws mainly on the use of dykes (for flood management) and nourishment (to preserve beaches). These costs have been documented globally by IPCC CZMS (1990) and Hoozemans et al. (1993), based on the global experience of Delft Hydraulics (now Deltares). Hence, the cost estimates are grounded in coastal engineering experience and are reasonably robust.

The UNFCCC study used the DIVA tool and focused on dyke construction and upgrade, and beach nourishment. The costs are derived from the earlier studies and the practical experience of Delft Hydraulics in coastal engineering projects around the world. Residual damages are also costed in terms of land values, depth-damage curves and the costs of relocating people. The computations are conducted on 12,148 coastal segments that collectively make up the world's coast, except for Antarctica (McFadden et al., 2007; Vafeidis et al., 2008). Hence, the UNFCCC study for coastal zones is based on extensive experience, and a good description of the adaptation measures, as well as a consideration of quite detailed data at the segment level. It is much more detailed than any earlier assessment tool.

5.4 Case studies and trends in coastal adaptation

Increasingly the world's coasts are 'engineered' and modified directly or indirectly by human agency. In some limited cases, present adaptation investment includes anticipating climate change, but in general this is not yet the case (e.g. Tol et al., 2008). Some of the limited cases of anticipatory adaptation are highlighted below. Even without climate change, growing populations and economic wealth in coastal areas suggests that substantial investment in coastal adaptation would be required through the 21st century.

Unlike adaptation in many other sectors, many coastal adaptation measures usually represent a collective government-led activity, reflecting that the coast is a shared resource (Klein et al., 2000). Hence, while some adaptation will need to be funded by private investment (e.g. port and harbour upgrades), much of the financing will be the responsibility of government. However, individual adaptation measures are also apparent. Insurance is a mechanism that helps private individuals gain resources to recover from disaster such as coastal flooding and is potentially an important response mechanism (Clark, 1998; Grossi and Muir-Wood, 2006). The availability of appropriate insurance varies greatly between coastal countries – it is unavailable in many developing countries and in mainland Europe (where the government is the insurer of last resort), while in the UK and USA it is the norm.

While there is significant interest in elaborating coastal adaptation measures and understanding their costs (e.g. UNFCCC, 1999; Klein et al., 2001; Bosello et al., 2007), hard numbers on investment in coastal adaptation are hard to identify as there is never a single 'ministry for coastal adaptation' with published accounts in any country. The reality is that coastal adaptation costs fall between government and the private sector, and different ministries are responsible for different aspects of the coastal adaptation process. For instance, in England and Wales, the major investment in coastal adaptation is in flood and erosion management, but the budget covers all flood and erosion management: i.e. flood management of all flood mechanisms, including inland flooding. Integrated coastal management in England and Wales is covered by a separated budget, and this investment is quite small compared to that invested in flood and erosion management.

Nicholls (2007) identified the following national/regional estimates of current investment:

- European Union: the total annual cost of coastal adaptation for erosion and flooding was an estimated €3200 million in 2001.

- England: the Flood and Coastal Management budget for coasts is roughly £250 million per year, and growing. New estimates show annual expenditure on all flood defence (rivers and coasts) rising from £575 million in 2011 to more than £1 billion in 2035, with the increase being primarily due to climate change (Environment Agency, 2009).

- Japan: the estimate of investment is 120–150 billion Yen per year from 2003 to 2006.

- The Netherlands: the corresponding estimate is $600–$1200 million (at 2006 prices), or 0.1–0.2% of GDP. This is expected to double or treble from 2020 to 2050 as the recent recommendations of the Delta Commission (2008) are implemented. This represents a

5

combination of climate change adaptation, looking 100 to 200 years into the future, and increasing safety to much higher levels (the risk of failure will be 1 in 100,000 in any year).

For individual projects, Nicholls (2007) identified the following costs:

- The Maldives: in the 'Safe Island' Projects for tsunamis the cost of reclamation and coastal protection, including harbour works for the Vilifushi project, was about $23 million.
- Venice: The MoSE Project to stop flooding of Venice costs roughly €4000 million.
- St Petersburg: the Flood Protection Barrier, Russia, was started in the 1980s, and then mothballed. It requires €440 million for completion.
- London: The Thames Estuary 2100 Project is investing £15 million on appraising the flood management options for London, including building a completely new downstream barrier. While nothing has been decided, costs of £4–6 billion for the 21st century have been mentioned for upgrade, while the sum of £10–20 billion has been mentioned for a new downstream barrier, which would be the response to a large rise in sea level (several metres).

5.5 An appraisal of the UNFCCC coastal costs

Here we reflect on the UNFCCC costs, and learn lessons on how future studies could be improved. The messages are generic and could be applied across most of the published studies on adaptation costs.

The costs reported by the UNFCCC for coastal adaptation appear reasonable as a snapshot cost for 2030 (see Nicholls, 2007). The costs are sensitive to the timeframe and sea-level-rise scenarios considered. If we assume that all flood defences anticipate conditions in 100 years, the annual costs in 2030 (assuming the investment occurs over 50 years) could be as high as $13 billion/year. If we assume a sea-level rise following current UK (DEFRA, 2006) or Dutch practice (Delta Commission, 2008), the costs would be raised significantly. Based on a high-end sea-level-rise scenario, adaptation costs would be roughly doubled. Whatever scenarios are considered, the costs are also likely to grow over the 21st century, following the expected acceleration in sea-level rise.

A potential deficiency of the UNFCCC study is the lack of consideration of other aspects of climate change such as more intense storms. The likelihood of this change is controversial (Meehl et al., 2007) but, if it did occur, the necessary adaptation costs (and residual damage costs) would rise substantially. While somewhat speculative, adaptation costs comparable to sea-level rise could occur. When combined with the uncertainties about sea-level rise already discussed, adaptation costs three times greater than those reported in the UNFCCC study are not implausible.

Another deficiency is the focus on the incremental adaptation costs, with less consideration of the residual damages, and no consideration of the adaptation deficit. Policy-makers need to be aware that, even if they invest the sums suggested in the UNFCCC study, serious damage could still occur due to the adaptation deficit. It is quite possible that the costs of addressing the adaptation deficit will exceed the costs of climate change as reported in the UNFCCC study ($11 billion/year), and this may be necessary to achieve successful adaptation. A key point is that the adaptation deficit for coastal areas is not well understood and requires further consideration: first at a conceptual level as to its definition and meaning, and then in a more quantitative sense. The issue of residual damage is also important as adaptation does not maintain an unchanging coast – the coast will change, and potential coastal habitats will be progressively degraded. The adaptation costs of addressing these issues will be higher, and some residual damage will remain.

Lastly, there are other profound changes occurring in coastal zones which, while not due to climate change, will interact adversely with climate change (Nicholls et al., 2007a). For example, many coastal cities are subsiding, causing a relative rise in sea level. Coastal and city managers will want to address all the causes of hazards, and not just those due to climate change.

In conclusion, the costing of adaptation in coastal areas is on a sounder basis than adaptation costs in many other sectors, as these studies are based on empirical evidence of coastal engineering and also case studies of response to relative sea-level rise (in subsiding megacities). However, the costs remain uncertain, and much higher costs than reported in the UNFCCC study are feasible if climate change is significant: this uncertainty should be assessed in future studies, including with multiple timeframes. Future assessments of coastal adaptation should also consider: (i) the full range of residual risk, (ii) the implications of the adaptation deficit, and (iii) wider coastal management issues such as subsiding coastal cities.

5

References

Bijlsma, L., Ehler, C.N., Klein, R.J.T., Kulshrestha, S.M., McLean, R.F., Mimura, N., Nicholls, R.J., Nurse, L.A., Perez Nieto, H., Turner, R.K. and Warrick, R.A. (1996) 'Coastal zones and small islands' in R.T. Watson, M.C. Zinyowera and R.H. Moss (eds) *Impacts, Adaptations and Mitigation of Climate Change: Scientific-Technical Analyses. (The Second Assessment Report of the Intergovernmental Panel on Climate Change, Working Group II)*. Cambridge University Press, Cambridge, 289–324.

Bosello, F., Kuik, O., Tol, R. and Watkiss, P. (2007) *Costs of Adaptation to Climate Change: A Review of Assessment Studies with a Focus on Methodologies Used*. Final Report to the European Environment Agency, 2 February, 128pp.

Burton, I. (2004) 'Climate change and the adaptation deficit' in: A. Fenech (ed.) *Climate Change: Building the Adaptive Capacity*. Papers from the International Conference on Adaptation Science, Management and Policy Options, Lijiang, Yunnan, China, 17–19 May, Meteorological Service of Canada, Environment Canada, 25–33.

Clark, M. (1998) 'Flood insurance as a management strategy for UK coastal resilience', *The Geographical Journal* 164(3), 333–343.

DEFRA (2006) *Flood and Coastal Defence Appraisal Guidance. FCDPAG3 Economic Appraisal. Supplementary Note to Operating Authorities – Climate Change Impacts*. October, Department of Environment, Food and Rural Affairs (DEFRA), London, 9pp (downloadable at: http://www.defra.gov.uk/environ/fcd/pubs/pag n/climatechangeupdate.pdf).

Delta Commission (2008) *Working Together with Water. A Living Land Builds for its Future*. Findings of the Deltacommissie 2008, The Hague (downloadable at http://www.deltacommissie.com/doc/deltareport_full.pdf).

Environment Agency (2009) *Investing for the Future: Flood and Coastal Risk Management in England: A Long-term Investment Strategy*. Environment Agency, Bristol (downloadable at http://publications.environment-agency.gov.uk/pdf/GEHO0609BQDF-E-E.pdf).

Fankhauser, S. (1995) 'Protection versus retreat: estimating the costs of sea-level rise', *Environment and Planning* A, 27, 299–319.

Grossi, P. and Muir-Wood, R. (2006) *Flood Risk in New Orleans: Implications for Future Management and Insurability*. Risk Management Solutions, Newark, NJ, 31pp (downloadable at http://www.rms.com/Publications/NO_FloodRisk.pdf).

Hinkel, J. and Klein, R.J.T. (2007) 'Integrating knowledge for assessing coastal vulnerability to climate change' in McFadden et al. (eds) *Managing Coastal Vulnerability*. Elsevier, Oxford, 61–77.

Hoozemans, F.M.J., Marchand, M. and Pennekamp, H.A. (1993), *A Global Vulnerability Analysis, Vulnerability Assessments for Population, Coastal Wetlands and Rice Production on a Global Scale* (2nd edn). Delft Hydraulics and Rijkswaterstaat, Delft and The Hague.

IPCC CZMS (1990) *Strategies for Adaptation to Sea Level Rise*. Report of the Coastal Zone Management Subgroup, Response Strategies Working Group of the Intergovernmental Panel on Climate Change, Ministry of Transport, Public Works and Water Management, The Hague.

Klein, R.J.T., Buckley, E.N., Aston, J., Nicholls, R.J., Ragoonaden, S., Capobianco, M., Mizutani, N. and Nunn, P.D. (2000) 'Coastal adaptation' in B. Metz, O.R. Davidson, J.W. Martens, S.N.M. Van Rooijen and L.L. Van Wie Mcgrory (eds) *Methodological and Technological Issues in Technology Transfer*. Special Report of the Intergovernmental Panel on Climate Change, Cambridge University Press, Cambridge, 349–372.

Klein, R.J.T., Nicholls, R.J., Ragoonaden, S., Capobianco, M., Aston, J. and Buckley, E.N., (2001) 'Technological options for adaptation to climate change in coastal zones', *Journal of Coastal Research* 17, 531–543.

McFadden, L., Nicholls, R.J., Vafeidis, A.T. and Tol, R.S.J. (2007) 'A methodology for modelling coastal space for global assessments', *Journal of Coastal Research* 23(4), 911–920.

Meehl, G.A., Stocker, T.F., Collins, W., Friedlingstein, P., Gaye, A., Gregory, J., Kitoh, A., Knutti, R., Murphy, J., Noda, A., Raper, S., Watterson, I., Weaver, A. and Zhao, Z.-C. (2007) 'Global climate projections', in S. Solomon, D. Qin and M. Manning (eds) *Climate Change 2007: The Physical Science Basis*. Contribution of Working Group I to the Fourth Assessment Report of the Intergovernmental Panel on Climate Change, Cambridge University Press, Cambridge, 747–845.

Narita, D., Tol, R.S.J. and Anthoff, D. (2009) *Economic Costs of Extratropical Storms under Climate Change: An Application of FUND*. Working Paper No. 274, ESRI, Dublin, 29pp.

Nicholls, R.J. (2004) 'Coastal flooding and wetland loss in the 21st century: changes under the SRES climate and socio-economic scenarios', *Global Environmental Change* 14, 69–86.

Nicholls, R.J. (2007) *Adaptation Options for Coastal Areas and Infrastructure: An Analysis for 2030*. Report to the United Nations Framework Convention on Climate Change, Bonn, 33pp.

Nicholls, R.J. and Klein, R.J.T. (2005) 'Climate change and coastal management on Europe's coast' in J.E. Vermaat, L. Ledoux, K. Turner, W. Salomons and L. Bouwer (eds) *Managing European Coasts: Past, Present and Future*. Springer, Environmental Science Monograph Series, 199–225.

Nicholls, R.J., Wong, P.P., Burkett, V., Codignotto, J., Hay, J., McLean, R., Ragoonaden, S. and Woodroffe, C.D. (2007a) 'Coastal systems and low-lying areas' in M.L. Parry, O.F. Canziani, J.P. Palutikof, P. van der Linden and C.E. Hanson (eds) *Climate Change 2007: Impacts, Adaptation and Vulnerability*. Contribution of Working Group II to the Fourth Assessment Report of the Intergovernmental Panel on Climate Change, Cambridge University Press, Cambridge, 315–357.

Nicholls R.J., Klein R.J.T., Tol R.S.J. (2007b) 'Managing coastal vulnerability and climate change: A national to global perspective' in McFadden et al. (eds) *Managing Coastal Vulnerability*. Elsevier, Oxford, 223–241.

Sugiyama, M., Nicholls, R.J. and Vafeidis, A. (2008) *Estimating the Economic Cost of Sea-level Rise*. Report No. 156, MIT Joint Program on the Science and Policy of Global Change, Boston.

Tol, R.S.J. (2007) 'The double trade-off between adaptation and mitigation for sea level rise: An application of FUND', *Mitigation and Adaptation Strategies for Global Change* 12, 741–753.

Tol, R.S.J., Klein, R.J.T. and Nicholls, R.J. (2008) 'Towards successful adaptation to sea level rise along Europe's coasts', *Journal of Coastal Research* 24, 432–450.

UNFCCC (United Nations Framework Convention on Climate Change) (1999) *Coastal Adaptation Technologies*. Technical Paper FCCC/TP/1999/1, United Nations Framework Convention on Climate Change Secretariat, Bonn, 49pp.

Vafeidis, A.T., Nicholls, R.J., Boot, G., Cox, J., Grashoff, P.S., Hinkel, J., Maatens, R., McFadden, L., Spencer, T. and Tol, R.S.J. (2008) 'A new global coastal database for impact and vulnerability analysis to sea-level rise', *Journal of Coastal Research* 24(4), 917–924.

Van Koningsveld, M., Mulder, J.P.M., Stive, M.J.F., Vandervalk, L. and Vanderweck, A.W. (2008) 'Living with sea-level rise and climate change: A case study of the Netherlands', *Journal of Coastal Research* 24, 367–379.

6

The costs of adapting infrastructure to climate change

David Satterthwaite and David Dodman

6

Summary

The fact that the UNFCCC estimates for the additional investment needed to adapt new infrastructure to climate-change risks in 2030 vary from $7.6 billion to $130 billion (UNFCCC, 2007) suggests a large uncertainty in the basis for making such estimates. The method used to make such estimates (based on 5–20% of the estimate of global investment in new infrastructure in 2030 that is vulnerable to the impact of climate change) is not appropriate for nations or regions with very inadequate or no provision for infrastructure and very inadequate investment flows into infrastructure. It omits the expensive investments in governance and technical capacity to operate this infrastructure effectively. Adding a small increment to infrastructure investment flows that are very inadequate is no basis for estimating the costs of adapting infrastructure. For low- and middle-income nations, what needs to be considered is the additional investment needed to provide the infrastructure for adaptation to climate-change risks in 2030, rather than the to adapt new infrastructure.

In the many nations where there is little or no government capacity to make the needed investments in building or adapting infrastructure, it is misleading to characterise 'the problem' as only a lack of funding. The UNFCCC estimates for the additional investment needed to adapt infrastructure to climate change risks in 2030 are as follows:

- For Africa, the estimates are between $22 million and $371 million. These figures are clearly absurdly low – only 0.2% of the global total – while much of this continent has little or no infrastructure in place to 'adapt' and very high levels of vulnerability to climate change.

- For 'developing Asia' the estimates are between $1.9 billion and $32.4 billion. 'Developing Asia' has close to 4 billion inhabitants and a high proportion of the world's population and settlements most at risk from climate change. As in Africa, much of this region has very large deficits in the needed protective infrastructure.

The estimated costs for removing the housing and infrastructure deficit in low- and middle-income countries overall is around $6.3 trillion in today's figures – so this implies a need for an annual average investment equivalent to $315 billion a year for 20 years to remove the infrastructure deficit. If the costs of adapting infrastructure to climate change in 2030 include a contribution to making good the deficit in basic infrastructure and building the governance capacity to invest in and maintain it (using an annual figure of one-twentieth of the total cost for this assigned to 2030, i.e. assuming a 20-year investment programme to remove the deficit), then:

- for all low- and middle-income nations, the costs of adapting infrastructure are around 137 times the low UNFCCC estimate and more than 8 times the high UNFCCC estimate

- for Africa, the costs are nearly 3000 times the lower UNFCCC estimate and over 170 times the higher UNFCC estimate

- for 'developing Asia' (including the Middle East), they are 114 times the lower UNFCCC estimate and 7 times the higher UNFCCC estimate

- for Latin America, they are around 100 times the low UNFCCC estimate and 6 times the high UNFCCC estimate.

6

The UNFCCC estimates are based on a very narrow consideration of 'infrastructure' which means that housing gets left out, although one of the most serious likely impacts of climate change is a large increase in damage to or destruction of housing. No consideration is given to the cost of developing the institutional/governance capacity to build and adapt needed infrastructure. In addition, no consideration is given to social infrastructure, despite its importance for reducing climate-change risk.

The UNFCCC estimates for the investments needed to adapt infrastructure to climate-change risks do not include the costs of the infrastructure that can dramatically reduce the impacts of extreme weather. For many extreme-weather events in urban and rural areas with large infrastructure deficits and poor-quality housing, good early-warning systems, measures taken just before the extreme-event and rapid and effective post-event responses (temporary accommodation, restoring access to services, supporting rapid return to settlements damaged and supporting rebuilding) greatly reduce the impacts on populations.

The method used by the UNFCCC for estimating the additional investment needed to adapt infrastructure includes no consideration of the infrastructural damage that cannot be prevented by adaptation, stemming both from conscious choice (locations/facilities/structures for which full protection is judged to be too costly) and from the incapacity on the part of those at risk and of those institutions responsible for reducing this risk (including local and national governments). In regard to residual damage, taking a conservative estimate, between 1996 and 2005, large extreme-weather disasters caused an average of more than 20,000 deaths a year and had economic costs averaging over $50 billion a year (most arising from flooding and windstorms). This is in effect the 'residual damage' from the inadequacies or impossibilities of protecting against extreme weather today. Climate change is likely to increase these costs considerably, even with adapted infrastructure.

6.1 The scope of 'adapting infrastructure'

The term infrastructure is generally considered to include 'the structural elements of an economy that facilitate the flow of goods and services between buyers and sellers' (*Macmillan Dictionary of Modern Economics*, 2008) and thus includes roads (of all sizes, from highways to streets and lanes) and bridges, railways, airports, ports, electric power systems, telecommunications, water, sewerage and drainage/waste-water management systems. The definition of infrastructure is sometimes broadened to include services which make economic and social activities possible – such as public transport, health care, education and emergency services (which collectively are sometimes termed social infrastructure). Broader definitions of infrastructure may even include housing; they may also include the 'institutional infrastructure' needed to operate, manage and invest in the economic and social infrastructure.

There are many grey areas for the definition – for instance, electricity grids may be unambiguously infrastructure, but what about power stations? Water mains are considered infrastructure, as are water treatment plans but what about the services that draw on them, such as water-vending, or private or community-managed wells used because of the inadequacies in provision for piped supplies? Roads may be considered infrastructure, but what about public transport? Ports, airports, roads and bridges are considered as infrastructure – but does the

6

'infrastructure' include only the physical facilities or also the institutions needed to manage them, such as navigation and traffic control for air and maritime transport?

The UNFCCC report (UNFCCC, 2007) is unclear about what it includes in regard to infrastructure. It implies that it recognises that buildings are part of infrastructure (see the comment on page 32 concerning 'Buildings and other infrastructure that might be damaged by the impacts of climate change') but its estimates for the additional investment needed to adapt infrastructure to climate change do not include housing. The report recognises that one important part of adaptation is the 'support activities that cope with climate-affected sectors or resources', including the infrastructure needed to support public health services (page 121), but then chooses not to take these into consideration.

Housing may get excluded on the assumption that it is the responsibility of housing owners to fund the adaptation (or climate-proofing) of their house or apartment. But if this is so, where do issues relating to the vulnerability of housing to climate-change impacts get considered? Housing might be considered to be a private responsibility, yet for informal/illegal settlements where over 800 million urban dwellers live in low- and middle-income nations (UN-Habitat, 2003a), the risks from extreme weather relate to their illegal status (which discourages investment in better-quality dwellings) and the refusal of governments to install infrastructure or provide services (Hardoy, Mitlin and Satterthwaite, 2001). The destruction of or damage to housing is one of the most common and most serious impacts of many extreme-weather events, especially in low-income and many middle-income nations. The damage to or loss of housing is usually concentrated among low-income groups and often also includes loss of possessions. Only a very small proportion of the population in low- and middle-income nations has insurance for this. Assessing the impacts of such events in terms of the value of property damaged or destroyed can be misleading; an event that is devastating to the lives of very large numbers of people (in deaths, injuries and loss of property) may have low economic impacts because of the low value assigned to the housing damaged or destroyed.

For infrastructure, adaptation costs should include the costs of limiting the impacts (as well as preventing them). For many extreme-weather events in urban and rural areas with large infrastructure deficits and poor-quality housing, good early-warning systems, measures taken just before the extreme event (e.g. reducing the impact of flooding by supporting populations to move temporarily to high ground or safe sites) and rapid and effective post-event responses (temporary accommodation, restoring access to services, supporting rapid return to settlements damaged and supporting rebuilding) greatly reduce the impacts on populations and their assets. Yet these measures might not be considered adaptation in that they are not limiting the damage done to infrastructure. The costs of building and maintaining this capacity to reduce the impacts of extreme-weather events is not included in figures for infrastructure investments and so is not considered in the UNFCCC estimates.

There is also the issue of infrastructural damage that cannot be prevented by adaptation, stemming both from conscious choice (locations/facilities/structures for which full protection is judged to be too costly) and from incapacity on the part of those at risk and those institutions responsible for reducing this risk (including local and national government). Thus, the UNFCCC estimates for infrastructure include consideration of a limited part of 'infrastructure' that does not include social infrastructure, disaster-response infrastructure, housing and the institutional infrastructure needed to build, maintain and adapt infrastructure.

6

The infrastructure deficit

All low-income nations and most middle-income nations have very large deficiencies in provision for infrastructure. For instance, high proportions of the urban populations in Africa and Asia and a significant proportion in Latin America live in homes and settlements with little or no infrastructure – that is, no all-weather roads, no drains, no piped water supplies and no provision for electricity (Hardoy, Mitlin and Satterthwaite, 2001; World Bank, 2006). Most urban centres have no sewers, including many with several million inhabitants (UN-Habitat, 2003b). An assessment in 2000 suggested that around 920 million urban dwellers in developing countries live in 'slums', most of which have inadequate or no provision for infrastructure (UN-Habitat, 2003a). The lack of provision or inadequacies in provision for protective infrastructure is perhaps the main reason for the very rapid increase in the number of flood and windstorm 'disasters' since the 1950s.

Reviewing data on disasters also gives some indications of the kinds of impacts that extreme-weather events can have on infrastructure – within the larger costs in terms of death, injury and economic disruption – and loss of livelihood for large numbers of people. For instance, the floods in Mozambique in 2000 destroyed or damaged roads, bridges and other forms of infrastructure, as well as causing several hundred deaths and the displacement of over 3 million people. Reviewing the 'disasters' registered on the international database between 1996 and 2005 shows not only thousands of people killed and tens of millions affected by floods and windstorms – but also damage worth hundreds of billions of dollars. For instance, in Asia, floods and windstorms between 1996 and 2005 caused over 70,000 deaths and economic loss of around $191 billion. A large part of these deaths and economic losses could be attributed to infrastructure deficiencies.

The UNFCCC report admits that there is a very large 'adaptation deficit' but then ignores this when coming up with estimates for infrastructure adaptation costs. For instance, the report notes:

'Evidence for the existence and size of the adaptation deficit can be seen in the mounting losses from extreme weather events such as floods, droughts, tropical cyclones, and other storms. These losses have been mounting at a very rapid rate over the last 50 years. This increase is likely to be mostly due to the expansion of human populations, socio-economic activities, real property, and infrastructure of all kinds into zones of high risk. Moreover, much of this property is built at a substandard level and does not conform even to minimal building codes and standards. This widespread failure to build enough weather resistance into existing and expanding human settlements is the main reason for the existence of an adaptation deficit. Real property and socio-economic activities are just not as climate-proof as they could and arguably should be. The evidence suggests strongly that the adaptation deficit continues to increase because losses from extreme events continue to increase. In other words, societies are becoming less well adapted to current climate. Such a process of development has been called "maladaptation".'
(UNFCCC, 2007, page 99)

Thus, the UNFCCC recognises that there is a very large climate-change adaptation deficit, much of which is an infrastructure and institutional infrastructure deficit, but it does not consider it appropriate to consider this in estimating adaptation costs for infrastructure.

6

6.2 The cost of addressing the infrastructure deficit

Detailed costings were done for five nations on the investments needed to meet the Millennium Development Goals between 2005 and 2015, and these came to $993–1047 per person (Sachs et al., 2005). Around half of this was for infrastructure (including water and sanitation, energy and roads). So if we assume that $500 per person is needed for infrastructure investments to meet the deficits and deficiencies in infrastructure provision to meet the Millennium Development Goals and apply this to the population of all low- and middle-income nations for 2010, the regional totals required over a ten-year period are:

- for Africa, $516 billion

- for Asia, $2000 billion (assuming that high-income nations in Asia are not included)

- for Latin America and the Caribbean, $297 billion.

But these are not estimates for eliminating all infrastructure (and other development) deficits. Many of the Millennium Development Goals are only for reducing the problem – for instance, halving the proportion of people without sustainable access to safe drinking water and basic sanitation by 2015. The goal for improving the lives of slum dwellers was only to reach 100 million slum dwellers by 2020, which would represent around 10% of the current slum population (and a much smaller percentage of the likely slum population in 2020). So to remove the infrastructure deficit completely is likely to cost far more; here we assume that it costs twice as much as the above estimates. Then we need to add to the costs of infrastructure an additional $700 billion that would be required to upgrade poor-quality housing and provide the infrastructure in urban areas for the expansion of the population (Table 6.1).

Table 6.1 Estimated costs for removing the housing and infrastructure deficit in low- and middle-income nations by 2030 (US$ billion)

Region	Costs 2005-2015	Costs 2015-2030	Costs of expanding housing & infrastructure for expanding urban populations	Total
Africa	516	516	200	1232
Low- and middle-income nations in Asia	2000	2000	350	4350
Latin America and the Caribbean	297	297	150	744
All low- and middle-income nations	**2813**	**2813**	**700**	**6326**

Source: Based on estimates in Sachs and the UN Millennium Project (2005) and UN Millennium Project (2005)

These estimates are also broadly in line with estimates in the 2009 report of the International Strategy for Disaster Reduction (ISDR, 2009), for the investments needed to reduce the deficit in disaster-risk avoidance and risk reduction. This suggests that several hundred billion dollars a year are required to address the underlying risk factors for disasters (including those relating to climate change). But this report quite rightly emphasises that the availability of funding is only a part of the solution, because solutions also depend on national and local governments with the competence, capacity and accountability to make the needed investments.

Of course, the above figures are based on 'guesstimates' and heroic assumptions – but they are at least likely to be within the right order of magnitude. Not all this funding would need to come from international development assistance agencies, as a proportion of the funding would come from taxes and other revenues from within the nations and from households. Here, it is difficult to estimate what proportion this would provide: for some nations (such as China), most will come from domestic sources; for many sub-Saharan African nations, most or all will have to come from external sources. But even if half the funding for this comes from the domestic resources of governments in low- and middle-income nations, this still means an external funding requirement of over $3 trillion to address current infrastructure deficits.

6.3 Challenging assumptions

UNFCCC estimates of the additional investment needed to adapt infrastructure to climate change risks in 2030 worldwide range from $7.6 billion to $130.1 billion (UNFCCC, 2007). At the lower level, this is similar to the estimated amounts required to adapt each of the four other sectors assessed (agriculture, forestry and fisheries; water supply; human health; and coastal zones); at the upper level it is approximately three times the total costs required for all these other sectors combined. Thus, the range of estimates for adapting infrastructure is very large; there is a seventeen-fold difference between the lower and the upper estimates, which must throw some doubt on the validity of the bases by which these estimates are made.

According to the UNFCCC, the estimate of investment needed to adapt new infrastructure to climate change in 2030 is calculated as follows:

1 the estimate for global investment in gross fixed capital formation in 2030 (which is at around three times the global investment in 2000): $22.27 trillion

2 multiplying by the proportion of this that is vulnerable to the impact of climate change (based on data for losses from weather disasters): 0.7% (Munich Re data) or 2.7% (ABI data): $153–650 billion a year

3 taking 5–20% of this total as the increase in capital costs needed for adaptation: $8–31 billion (Munich Re data) or $33–130 billion (ABI data).

This produces the figures shown in Table 6.2.

6

Table 6.2 UNFCCC figures for additional investments needed to adapt infrastructure to climate change risks in 2030

Region	Total (US$ billion)
Africa	0.022 – 0.37
Low- and middle-income nations in Asia (including Middle East)	2.0 – 33.5
Latin America and the Caribbean	0.4 – 6.9

Source: UNFCCC (2007), Table V–53, page 123

There are three weaknesses in this approach:

1 The estimate for global investment in 2030 is simply an extrapolation of data for 2000. Neither it nor the data for 2000 include any estimate of what investment flows are needed for infrastructure. They make no allowances for vast deficiencies in infrastructure provision. This can be seen most dramatically in the case of Africa which is projected to have only 2.2% of world investment but 17.6% of the world's population in 2030. The projected increases in global investment fixed capital formation will not of itself address these deficiencies.

2 (As the study notes), the Munich Re data are likely to substantially under-estimate damages from climate because only data from large events are included. But this may lead to a very large under-estimate as the cost of extreme-weather disasters that are not recorded by Munich Re or by official disaster statistics (see for instance ISDR, 2009).

3 It is not possible to estimate the cost of adapting infrastructure by adding a percentage to existing investments if there are no existing investments.

To date, calculations of the costs of adapting infrastructure to climate change have been predominantly top-down in approach, based on conditions, data and assumptions drawn from high-income nations. For example, UNFCCC estimates are based on the *additional investment required*, calculated as a percentage of the estimate for investment flows into climate-sensitive infrastructure. This generates absurdly low figures in some cases: according to the UNFCCC figures, as little as $22 million may be required for adapting infrastructure for all of Africa in 2030 (5% of the investment in infrastructure that is vulnerable to climate change). The same calculation suggests that the cost of adapting infrastructure in OECD North America in 2030 will be $3.7 billion. At the top end of the estimates (20% of ABI estimates), Africa is expected to require $371 million and OECD North America $63.7 billion. These figures therefore suggest that infrastructure in OECD North America will require 170 times more investment to adapt to climate change than will infrastructure in Africa.

The UNFCCC estimates for the additional investment needed to adapt infrastructure to climate change risks do not work for locations with the largest infrastructure deficits and with the least governance capacity to address these deficits. The figures for Africa are clearly nonsensical. For much of this continent, there is little or no infrastructure in place 'to adapt' and very high levels of vulnerability to climate change (Boko et al., 2007; Bicknell et al., 2009).

If you take the figures for the investments needed to remove the infrastructure deficit in low- and middle-income countries given in Table 6.1, and then assume that an additional 5% or 20% investment is needed to adapt to climate change, you get the estimates in Table 6.3. But note that the figures for the costs of adapting the infrastructure to climate change are valid only if the investments are made in removing the infrastructure deficit.

Table 6.3 Total estimated costs for removing the housing and infrastructure deficit by 2030 plus additional total costs for adapting to climate change

Region	Total costs (US$ billion) of: Removing the infrastructure deficit	Adapting to climate change (5%)	Adapting to climate change (20%)	Annual costs over 20 years (US$ billion) of Adapting to climate change (5%)	Adapting to climate change (20%)
Africa	1232	61.6	246.4	3.1	12.3
Low- and middle-income nations in Asia	4350	217.5	870	10.9	43.5
Latin America and the Caribbean	744	37.2	148.8	1.9	7.4
All low- and middle-income nations	6326	316.3	1265.2	15.9	63.2

Note: this table shows the total estimated costs for removing the housing and infrastructure deficit in low- and middle-income nations by 2030, the additional total costs for adapting to climate change and the annual costs, assuming investments are made over 20 years

6

Table 6.4 Comparing UNFCCC figures and our alternative figures for the additional investment needed to adapt infrastructure to climate change risks in 2030 (US$ billion)

Region	1) UNFCCC (as shown in Table 6.2)	2) Our estimates for the annual investment needed to contribute to removing the infrastructure deficit*	3) Our estimates for the additional costs needed to adapt infrastructure	4) Our estimates for the annual investment needed to remove the infrastructure deficit (over 20 years) & adapt infrastructure (column 2+column 3)
Africa	0.022 – 0.37	61.6	3.1 – 12.3	64.7 – 73.9
Low- and middle-income nations in Asia (including Middle East)	2.0 – 33.5	217.5	10.9 – 43.5	228.4 – 261.0
Latin America and the Caribbean	0.4 – 6.9	37.2	1.9 – 7.4	39.1 – 44.6
All low- and middle-income nations	2.422 – 40.8	316.3	15.9 – 63.2	332.2 – 379.5

* This is the total needed to remove the infrastructure deficit in the second column in Table 6.3 divided by the 20 years over which this annual investment is needed. Of course, if this investment began in 2010, 2030 would be the first year for which this was not needed, because the investments would have been made for 2010 to 2029

Source: UNFCCC figures from UNFCCC (2007)

Column 4 in Table 6.4 shows our estimates for the costs of adapting infrastructure to climate change in 2030. This includes a contribution to making good the deficit in basic infrastructure and building the governance capacity to invest in and maintain it (one-twentieth of the total cost for this, as this is spread over the period 2010 to 2030) and a 5–20% addition to cover the extra costs of addressing climate change. Comparing this with the UNFCCC figures and taking the lower of the two estimates in column 4, the needed investment for Africa in 2030 would be nearly 3000 times the lower UNFCCC estimate and over 170 times the higher UNFCC estimate. For Latin America, the needed investment in 2030 would be around 100 times the low UNFCCC estimate and nearly 6 times the high UNFCCC estimate.

The UNFCCC estimates for the additional investment needed to adapt infrastructure to climate-change risks in 2030 in what it terms 'developing Asia' are between $1.9 billion and $32.4 billion. If the estimates for the Middle East are added to this, this comes to $2–33.5 billion. 'Developing Asia' has close to 4 billion inhabitants and high proportion of the population and cities most at risk from climate change. As in Africa, much of this region has very large deficits in the needed protective infrastructure. If the costs of adapting infrastructure to climate change include making good the deficit in basic infrastructure and building the governance capacity to invest in and maintain needed infrastructure, for 'developing Asia' (including the Middle East), the needed investment in 2030 would be 114 times the lower UNFCCC estimate and 7 times the higher UNFCCC estimate. 'Developing Asia' has more than half the world's large cities (those with 1 million or more inhabitants), including nine cities with more than 10 million inhabitants (United Nations, 2008). Just the cost of producing the infrastructure needed to protect some of Asia's large coastal cities particularly at risk from cyclones/typhoons and/or sea-level rise is likely to be several billion dollars by 2030.

The whole basis for estimating the investment needed to adapt infrastructure is therefore a false premise – that this investment can be costed by applying a small increment to existing investment flows into infrastructure that is climate sensitive. This takes no account of the very large deficits in basic infrastructure that are evident in all low-income and most middle-income nations. This leads to the entirely wrong conclusion that most of the investment needed for climate-change adaptation for infrastructure is required in high-income nations, rather than in low- and middle-income nations. It also ends up showing very small sums needed for Africa and other places where there are very low or inadequate investment flows into infrastructure, and where many of the nations most at risk from climate change are located.

Three other assumptions need to be questioned:

1 That the availability of funding from international agencies is the 'solution' for adaptation (and perhaps for the adaptation deficit too which is mentioned by UNFCCC but not included when calculating costs). In much of Africa and Asia and parts of Latin America and the Caribbean, local governments are weak, ineffective and unaccountable to local populations so their capacity to design and implement appropriate adaptation strategies that serve those most at risk from climate change must be in doubt. This is most obvious in the nations often termed 'failed states' but also evident in many other nations. External funding agencies have not proved very effective in addressing this – or even in knowing how to address it.

2 That 'adaptation' and 'development' can be kept separate. On the ground, climate-change impacts are exacerbating non-climate-change impacts, and addressing both is inhibited by institutional/governance failures. It is difficult if not impossible to separate what proportion of extreme-weather damages or water shortages in any locality is caused by climate change. Also, so much of the adaptation deficit for housing and infrastructure is also a development deficit.

3 That National Adaptation Programmes of Action (NAPAs) give us an idea of adaptation costs. The focus of most of the NAPAs is a very small part of what many nations will need, so these plans are not a good basis for assessing adaptation costs.

Example: coastal infrastructure in Tuvalu

One of the projects identified by the NAPA of Tuvalu is coastal infrastructure development to protect the shoreline from erosion – a problem regardless of climate change (and so an existing development need), but one exacerbated by climate change (so also an additional cost). Distinguishing between 'additional' and 'baseline' adaptation needs on the ground has proved extremely difficult. Further, being a poor country, Tuvalu cannot afford to meet the costs of baseline infrastructure. Thus, the offer to fund the 'top section' of the infrastructure required,

6

to respond to 'additional' impacts of climate change, is absurd in light of the fact that co-financing to pay for the lower section cannot be found. The project is currently in limbo while co-financing is sought (Ayers and Huq, 2008).

The focus on the 'additional funding' needed for adaptation also ignores the role of effective local institutions, including municipal authorities and civil society organisations. In some cases, large amounts of money could be spent with little effect; in other cases, relatively small (and often locally generated) investments could yield significant results. For instance, the construction and maintenance of city-wide urban sanitation and drainage systems in Karachi was done through partnerships between community/resident organisations (who installed the systems within communities) and the government (who installed the trunk sewers and drains into which these feed). Supported by a Karachi-based local NGO, the Orangi Pilot Project Research and Training Institute based in Karachi, this has proved very effective and fundable with local resources. In contrast, a design prepared by international consultants and to be funded by the Asian Development Bank, would have been four to five times the price, would have needed considerable external funding – and probably would not have worked (Hasan, 1999; 2006).

Thus, the result of using an inappropriate method for estimating the additional investments needed for adapting infrastructure for low- and middle-income nations leads to vast underestimates for their funding needs for infrastructure. It also leads to the finding that infrastructure adaptation costs will be higher in high-income nations than in low- and middle-income nations. Yet, most of the world's population and settlements that are most at risk from climate change and most lacking in infrastructure are in low- and middle-income nations.

6.4 'Residual damage'

Even if the investments needed to remove the infrastructure deficits, plus the needed adjustments for climate-change risks, are made, this does not prevent all climate-change costs. Table 6.5 gives estimates for the mortality and economic damage done by the 'large' disasters that are recorded in international disaster statistics. These are known to understate the impact of disasters, both through the disasters that they do not record and because of the very large number of disasters that do not meet the official criteria for inclusion as a disaster.[1] Nevertheless, the figures included in Table 6.5 show that floods and windstorms caused damage worth over $500 billion and over 150,000 deaths in one decade. They also affected over 1.6 billion people. In 2008 alone, there were 140,848 deaths from storms although most of these were from the impact of Cyclone Nargis on Myanmar (EM-DAT). These are indications of the cost of not having the infrastructure in place to prevent damage and loss of life from windstorms and floods. This is not to claim that this is a climate-change cost – but it is an indication of the costs that have not been prevented from extreme-weather events whose frequency and/or intensity climate change is likely to increase. Note too the mortality and number of people affected by avalanches and landslides and by extreme temperatures in Table 6.5; these are also extreme events to which climate change will contribute.

Table 6.5 Global extent and impacts of certain disasters by hazard type, total 1996–2005

	Number of events	Mortality	People affected (thousand)	Economic damage (US$ million, 2005 prices)
Avalanches/landslides	191	7864	1801	1382
Earthquakes, tsunamis	297	391,610	41,562	113,181
Extreme temperatures	168	60,249	5703	16,197
Floods	1310	90,237	1,292,989	208,434
Volcanic eruptions	50	262	940	59
Windstorms	917	62,410	326,252	319,208
Industrial accidents	505	13,962	1372	13,879
Miscellaneous accidents	461	15,757	400	2541
Transport accidents	2035	69,636	89	960

Source: EM-DAT

At present, each year, damage worth tens of billions of dollars is done by extreme weather in high-income nations that have their populations served by protective infrastructure and good-quality buildings. So this is an indication of costs that have not been avoided by 'adaptation' to extreme weather, even in nations with high adaptive capacity. There has been a very large increase in the number of natural disasters since 1950 and most of this increase is from the increase in extreme-weather events (EM-DAT), whose intensity and frequency climate change is likely to increase.

6.5 Towards new methodologies

There is a need for careful case studies of what adaptation would involve in particular locations, and what component of this is from infrastructure. These case studies would need to consider the infrastructure deficit, and the institutional/governance underpinning necessary for addressing the deficit and climate-proofing all new and existing infrastructure. This could lead to a better idea of the kind of funding needed for adapting infrastructure to climate-change risks, and then to some thoughtful discussion of what this implies for adaptation costs and adaptation funding in general. It would take only a few such studies of major cities particularly at risk from climate change and with large infrastructure deficits to show that the UNFCCC estimates for Africa and for 'developing Asia' are far too low. It is also likely that studies of major cities in Latin America and the Middle East at high risk from climate change would show the UNFCCC estimates for these regions to be far too low.

Even with a growing number of careful location-based estimates for costs, however, it will be difficult to extrapolate these to figures for whole regions. This is because:

- There are very large differences in contexts (risks and vulnerability), including the scale of infrastructure deficits and the extent of local governance failures. In most of the locations with the largest infrastructure deficits and governance failures, much of the data needed to assess such costs are not there.

6

- There are very large differences in costs. In the south-east of England, an effective package of passive measures to control overheating in the summer is expected to cost around £16,000 for a typical unadapted house (Three Regions Climate Change Group, 2008) – enough to build 15 to 20 houses in many Asian and African urban centres.
- The 'moving target' of urbanisation is likely to result in almost all growth in the world's population in the next few years to be in urban areas in low- and middle-income nations.
- Many of the costs in adapting cities – particularly in upgrading housing stock – will be borne by private individuals, and so are even more difficult to account for than are public costs. Estimates based on the costs of adapting infrastructure are thus certainly not the 'total costs of adaptation'.

Example: London

London's climate-change adaptation strategy assesses the climate-change impacts that will be felt as a result of flooding, drought and overheating, and assesses the implications of this for health, the environment, the economy and infrastructure. This strategy includes considerable detail on the different adaptation needs (e.g. from water supply systems to flood barriers to damage to train lines) – but no 'bottom-up' costings of these within the main adaptation document (Three Regions Climate Change Group, 2008).

Example: Cape Town

One of the most detailed 'Municipal Adaptation Plans' in Africa has been undertaken for Cape Town. This assesses issues of fresh-water availability, stormwater drainage, fires and risks for coastal zones, but it does not have any figures attached to it (Mukheibir and Ziervogel, 2007).

Example: Durban

The municipal government of Durban has been considering issues of climate protection for several years (this case study draws from Roberts, 2008). This included the Climate Future for Durban Programme initiated in 2004 with the Council for Scientific and Industrial Research (a parastatal research organisation) to consider the implications of global and regional climate-change science for Durban, and the development of a Headline Climate Change Adaptation Strategy for the city to highlight how key sectors within the municipality should begin responding to unavoidable climate change and incorporating climate change into long-term city planning. The municipality and CSIR are working with the Tyndall Centre for Climate Change Research (in the UK) to develop a model that will enable the simulation, evaluation and comparison of strategic urban development plans within the context of climate change. The Headline Climate Change Adaptation Strategy includes predicted changes and likely effects of climate change, 2070–2100 for infrastructure, population health, food security and agriculture, water, tourism/business and biodiversity/coastal zone. This makes clear the relevance of climate-change issues for virtually all departments and agencies within the municipal government. This work emphasises how each department of local government needs to consider the effects of climate change within its departmental responsibilities. Only when this happens, and when climate science can provide each locality with a stronger idea of climate-change impacts and how these will change over time, will detailed costings for infrastructure adaptation become possible.

6

Example: Gulf of Mexico

A detailed assessment of the possibilities and constraints on needed climate change adaptation in the coastal states around the Gulf of Mexico (home to more than 55 million people) hardly mentioned monetary costs as it described in detail the weakness in the local governments as they lack the financial, human and technical capacity to fulfil responsibilities for urban development plans, zoning and land-use management (including granting permits for construction). The profits that can be generated by changes in land-use designation in and around urban centres, and in contracts for public works or services, make corruption in local government difficult to control, within both Mexico and the USA. The coastal states have a heavy concentration of economic activities – petroleum production, fisheries, agriculture, forestry and tourism. The Gulf concentrates a high proportion of US offshore oil production and of Mexico's total oil production, and the oil and gas industry supports a very large complement of land-based companies and facilities including chemical production, oil-field equipment dealers, cement suppliers, caterers, divers, platform fabrication yards and shipyards. There are many examples of local governments in the USA not fulfilling their responsibilities on risk reduction and allowing buildings and enterprises in high-risk zones. There are also examples of perverse public policies and subsidies that act to increase development in high-risk zones (Levina et al., 2007).

Example: Massawa, Eritrea

A rise in sea level of 0.5m would submerge infrastructure and other economic installations in Massawa to a total value of over $250 million. This is more than ten times the lowest estimate given by the UNFCCC for adapting all infrastructure in all of Africa (State of Eritrea, 2001).

Example: Iloilo, Philippines

A flood-control infrastructure project is currently being constructed at a cost of approximately $100 million, based on 1996 estimates of a 20-year return period. According to the engineers working on this project, increasing the specifications to withstand a 50-year return flood (even based on 1996 figures) would have resulted in a very considerable increase in costs (Dodman et al., forthcoming).

The examples from two small cities in Eritrea and the Philippines illustrate how much even relatively small-scale interventions will cost. There are not only hundreds of large cities but also tens of thousands of small urban centres in 'developing Asia', Africa, the Middle East and Latin America and the Caribbean with serious infrastructure deficits. There are also billions of rural dwellers in these regions that lack homes and infrastructure that protects them and their livelihoods from climate-change impacts. This suggests that the UNFCCC figures for infrastructure investment are far too low – for Africa, possibly by two or three orders of magnitude. Without a larger number of detailed local studies, however, there is no basis for suggesting more realistic adaptation costs.

6

6.6 Directions for policy: addressing the real costs of adapting infrastructure

- Reduced emphasis on 'global costs', and much more on supporting careful local estimations and building regional and global estimates from these.

- Increased attention to understanding the 'infrastructure deficit' (and beginning to cost it), and the 'institutional/governance' deficits that often underpin this. It makes no sense to discuss adaptation for infrastructure and for housing without this understanding.

- Creating frameworks to generate effective and comparable cost estimates 'bottom-up' within cities and countries through NAPAs; NAPAs need LAPAs (local adaptation plans of action) to underpin them and their cost estimates.

- Consider where housing fits into the discussions of infrastructure; the high levels of vulnerability of large parts of the housing stock are related to inadequate or no provision of infrastructure.

1 To be entered into the EM-DAT database, at least one of the following criteria has to be fulfilled: 10 or more people reported killed; 100 people reported affected; a call for international assistance; and/or declaration of a state of emergency.

References

Ayers, J., and S. Huq (2008), 'Supporting adaptation through development: What role for ODA?' Paper presented at DSA Annual Conference, November 2008, *Development's Invisible Hands: Development Futures in a Changing Climate*.

Bicknell, Jane, David Dodman and David Satterthwaite (editors) (2009), *Adapting Cities to Climate Change: Understanding and Addressing the Development Challenges*, Earthscan, London.

Boko, M., Niang, I., Nyong, A., Vogel, C., Githeko, A., Medany, M., Osman-Elasha, B., Tabo, R. and Yanda, P. (2007), *Africa. Climate Change 2007: Impacts, Adaptation and Vulnerability. Contribution of Working Group II to the Fourth Assessment Report of the Intergovernmental Panel on Climate Change*. Parry, M., Canziani, O., Palutikof, J., van der Linden, P. and Hanson, C. (eds). Cambridge University Press, Cambridge, 433–467.

Dodman, D., Mitlin, D. and Rayos, Co. J. (forthcoming), 'Victims to victors, disasters to opportunities: community driven responses to climate change in the Philippines'. Accepted (subject to revisions) for *International Development Planning Review*.

EM-DAT, the OFDA/CRED International Disaster Database, www.emdat.be, Catholic University of Louvain, Brussels.

Hardoy, Jorge E., Diana Mitlin and David Satterthwaite (2001), *Environmental Problems in an Urbanizing World: Finding Solutions for Cities in Africa, Asia and Latin America*, Earthscan, London, 448 pages.

Hasan, Arif (1999), *Understanding Karachi: Planning and Reform for the Future*, City Press, Karachi, 171 pages.

Hasan, Arif (2006), 'Orangi Pilot Project; the expansion of work beyond Orangi and the mapping of informal settlements and infrastructure', *Environment and Urbanization*, Vol. 18, No. 2, pages 451–480.

ISDR (International Strategy for Disaster Reduction) (2009), *Risk and Poverty in a Changing Climate: Global Assessment Report on Disaster Risk Reduction*. United Nations, Geneva.

Levina, Ellina, John S. Jacob, Luis E. Ramos and Ivonne Ortiz (2007), *Policy frameworks for Adaptation to Climate Change in Coastal Zones: The Case of the Gulf of Mexico*, Paper prepared for the OECD and International Energy Agency, 68 pages.

Macmillan Dictionary of Modern Economics (2008) quoted in http://www.med.govt.nz/templates/MultipageDocumentPage____9202.aspx

Mukheibir, Pierre and Gina Ziervogel (2007), 'Developing a Municipal Adaptation Plan (MAP) for climate change: the city of Cape Town', *Environment and Urbanization*, Vol. 19, No. 1, pages 143–158.

Roberts, D. (2008), 'Thinking globally, acting locally – institutionalizing climate change at the local government level in Durban, South Africa', *Environment and Urbanization*, Vol. 20, No. 2, pages 521–537.

Sachs, Jeffrey D. and the UN Millennium Project (2005), *Investing in Development; A Practical Plan to Achieve the Millennium Development Goals*, Earthscan, London and Sterling, VA, 224 pages.

State of Eritrea (2001), *Eritrea's Initial National Communication under the United Nations Framework Convention on Climate Change*.

Three Regions Climate Change Group (2008), *Your Home in a Changing Climate: Retrofitting Existing Homes for Climate Change Impacts*. Greater London Authority.

UN Millennium Project (2005), *A Home in the City*, The report of the Millennium Project Taskforce on Improving the Lives of Slum Dwellers, Earthscan, London and Sterling, VA, 175 pages.

UNFCCC (2007), *Investment and Financial Flows to Address Climate Change*, United Nations Framework Convention on Climate Change, Bonn, 270 pages.

UN-Habitat (2003a), *The Challenge of Slums: Global Report on Human Settlements* 2003, Earthscan, London.

UN-Habitat (2003b), *Water and Sanitation in the World's Cities; Local Action for Global Goals*, Earthscan, London, 274 pages.

United Nations (2008), *World Urbanization Prospects: the 2007 Revision*, CD-ROM edition, data in digital form (POP/DB/WUP/Rev.2007), United Nations, Department of Economic and Social Affairs, Population Division, New York.

World Bank (2006), *2006 World Development Indicators Online*, The World Bank, Washington DC.

7

Costing adaptation for natural ecosystems

Pam Berry

7

Summary

- The UNFCCC study is the first and only one attempting to estimate costs of adaptation in natural ecosystems.

- The ability to cost this fully is strictly limited due to lack of data on the effectiveness of proposed adaptation measures, the size of the adaptation deficit and knowledge on residual damages. Important assumptions and extrapolations, therefore, are inevitable.

- The UNFCCC costs, based on enhancing the global terrestrial protected areas network, suggest that an annual increase in expenditure of $12–22 billion is needed. This does not include marine protected area costs, or adaptation in the wider landscape. Thus it is an under-estimate of the full costs of adaptation.

7.1 Introduction

Climate change is projected to have far-reaching impacts on ecosystems and their species (Fischlin et al., 2007), but although species have a certain capacity to adapt autonomously, this is often limited and can be adversely affected by actions undertaken in other sectors. Human intervention, therefore, will be necessary to enhance this adaptation and to maximise the security of ecosystems and the services which they provide. This chapter analyses the various issues, assumptions and methods surrounding existing studies of global costings of adaptation, and, in particular, critically evaluates the UNFCCC study on natural ecosystems. These were not included in the overall global costs, owing to the nature and level of the assumptions and the uncertainties involved in the calculations, as detailed below.

7.2 Issues and assumptions

Definitions

The various studies on costing adaptation have defined and dealt with ecosystems in different ways. The World Bank (2006) when examining climatic risks mentions natural resources in terms of recognising the ecosystem services provided. The Stern Review (2006) refers to natural ecosystems in the chapter on adaptation, while the UNDP report (2007) has one passing reference to the vulnerability of ecosystems in its chapter on adaptation. The Oxfam report refers to impacts on natural resources and ecosystems (Oxfam, 2007, page 11) and recognises the need to protect ecosystems and existing infrastructure 'so that they are resilient to the coming stress from climate change' (Oxfam, 2007, page 15). The UNFCCC report (2007) is the only study to deal explicitly with 'natural ecosystems'[1] and this is similar to Tol (2002) who includes 'natural (unmanaged) ecosystems' in his estimates of damage costs of climate change. Thus, while all the studies mention ecosystems, most do not define them and there is an issue as to whether ecosystem services are or should be included in any assessment of the costs of adaptation.

7

What is costed

The World Bank (2006), Stern Review (2006), Oxfam (2007) and UNDP (2007), while mentioning ecosystems briefly in the text, are not explicit about how they have been incorporated into their costing of adaptation, and the UNFCCC was the only study to provide explicit details of how the costs of adaptation were calculated (Berry, 2007). This raises the issue of costing adaptation and whether it is purely associated with the existence of current and future ecosystems and their component biodiversity or whether ecosystem services are included, either within natural ecosystems or separate from them. This is important if double counting is to be avoided. Indeed this is very difficult as any land-use change as a consequence of adaptation (or mitigation) will affect natural ecosystems, but they could interact synergistically or antagonistically (Berry et al., 2008; Paterson et al., 2008). For example, wetland re-creation as a flood management adaptation also would be an appropriate adaptation measure for ecosystems, whereas hard coastal defences are an option for adaptation to coastal flooding, but they constrain ecosystems' autonomous adaptation. A further issue is that none of the studies, except UNFCCC, explicitly considered marine ecosystems.

Adaptation strategies

Costing adaptation for natural ecosystems, is based on the premise that autonomous adaptation will be inadequate (Fischlin et al., 2007) and thus human, planned adaptation is necessary to avoid the projected negative impacts on biodiversity (and ecosystem services). The possible costs of the *impacts* of climate change for natural ecosystems have been assessed globally by several studies, including Nordhaus and Boyer (2000) and Tol (2002), but these are often based on broad assumptions in the absence of relevant data. Given the difficulty of establishing the costs of impacts, no studies have established the effectiveness and costs of adaptation options in ecosystems, although some qualitative adaptation opportunities and costs were identified in the IPCC report for certain ecosystems and regions (IPCC, 2007).

An important issue for costing adaptation is that although a number of adaptation options have been suggested for natural ecosystems which could be applied singly or in combination (e.g. Pöyry and Toivonen, 2005; Convention on Biological Diversity; Hopkins et al., 2007), they are based largely on ecological theory and have not been tested for their effectiveness in practice (Brooker and Young, 2006). Also, many of the options are local and context specific, which makes costing of global adaptation difficult. Many of the adaptation options identified, however, should also form part of current good conservation practice, thus the UNFCCC based its costings on studies by James et al. (2001) and Balmford et al. (2003) of bringing the terrestrial and marine protected area networks (PAN) up to a minimum standard, and of enhancing them. This could be considered as addressing a current deficit, but much of this deficit is due to habitat destruction and fragmentation and unfavourable land management practices, rather than direct climate change. A number of assumptions, approximations and extrapolations were involved in the costings, which will be highlighted in Section 7.3 below.

7

The level of adaptation

The UNFCCC assessed adaptation under two scenarios – 'business as usual' (BAU) and mitigation. Both of these had no explicit level of adaptation for climate change to be achieved, but relied on the assumption that an adequate network would provide at least the first step in providing the necessary adaptation to climate change. Thus the extent of residual damage was not estimated and indeed is unknown at the global level. The mitigation scenario used the World Conservation Union's suggestion that at least 10% of the land area of each nation or ecosystem be set aside for conservation (IUCN, 1993) and the costs of achieving this were calculated. For the BAU scenario it was assumed that due to a greater climate change, PAN would need strengthening, and the costs were calculated by increasing the percentage of higher protection levels of IUCN categories within this 10%.

This level of adaptation was thought to be inadequate given the projected changes in biodiversity and, more particularly, it would not ensure the maintenance of ecological and evolutionary processes, which are an important part of autonomous adaptation (James et al., 2001). Also, reserves need to be seen in a wider landscape context, as species may be (partly) dependent on the surrounding land and its use may add to the pressures on PAN. Thus they explored the additional costs involved in establishing a wider matrix for conservation.

Limitations

The greatest limitation of the above for application to costing climate change adaptation in natural ecosystems is that the work was focused on PAN, which forms only part of the range of adaptation options. Also, there is no evidence that this will be sufficient for the projected climate change impacts, and there is no information on the possible magnitude of the shortfall. In addition, the antagonistic cross-sectoral interactions of adaptation actions mentioned above, or non-climate pressures on ecosystems may mean that adaptation requirements will be greater than those due to climate alone.

7.3 Method

The UNFCCC methodology consisted of estimating: the current global expenditure on conservation in the form of protected areas (PAs); the shortfall in the protected area network (PAN); the level of additional expenditure needed for PAN to be adequate for climate change adaptation; and costing adaptation outside the protected area network. Each of these steps is considered in this section.

Current global expenditure on protected areas

The UNFCCC figures on current expenditure were derived from James et al. (2001). They used a combination of data on PA expenditure from World Conservation Monitoring Centre (WCMC) questionnaire surveys of PA agencies in 1993 and 1995. Responses were received from 123 conservation agencies in 108 countries that manage more than 28% of the global PA system. Where possible data included national government budgets for both recurrent management expenses and capital outlays, plus foreign and non-governmental assistance. Budgetary

7

information for another 2.55 million km² was obtained from the grey literature, so the figures included 47.5% of the global PA system. The global estimate of expenditures on PAs was calculated by dividing the world into 10 economically and ecologically similar regions. Within each region, the available data for countries were extrapolated to obtain an estimate for the entire region and global cost estimates were the sum of the 10 regional estimates. There is no indication of the quality of the input data or detail on the extrapolation from countries to a region.

There is some debate about the source of funding for current expenditure on PAs, with Pearce and Palmer (2001) suggesting that the private sector probably accounts for less than half of the total, while Emerton et al. (2006), citing Lapham and Livermore (2003), suggest that non-governmental and private sector funding is now providing the bulk of PA finance. The GEF Secretariat is certainly trying to increase engagement with the private sector and while a large number of projects are funded in the name of biodiversity it is difficult to get an overall clear picture of the nature of these projects and the proportion of public and private funds invested.

Shortfalls in the protected area network

Shortfalls in the existing reserve network also were derived from James et al. (2001). They used the WCMC survey (James et al. 1999) with data from PA managers in 52 developing countries and from 14 developed countries. These estimates formed the basis of an extrapolation of PA budget shortfalls for the developing countries and for Europe. The data were insufficient to make similar estimates for the remaining developed countries, so it was assumed that their budgetary shortfalls averaged 10% of actual expenditure (James et al., 2001).

Given the inadequacy of the existing PAN to meet current conservation objectives, the cost of purchasing land to expand the network for ecological representation, their future management and the scale of compensation required to meet the opportunity costs incurred by local people living in or near reserves were also calculated (James et al., 2001). In order to estimate PA network expansion, those authors adopted the World Conservation Union's suggestion that at least 10% of the land area of each nation or ecosystem be set aside for conservation (IUCN, 1993). The costs of this were added to the current and shortfall sums to obtain the annual cost of adequate biodiversity conservation within a global reserve system. The shortfall in current conservation expenditure was supplemented by two other sources (Balmford et al., 2002 and Bruner et al., 2004).

Balmford et al. (2002) used data on the recurrent management costs per unit area of effective terrestrial field-based conservation programmes (57 sites) from James et al. (1999), from correspondence with local experts (21 sites), from the published and unpublished literature (20 sites), and from the World Wide Web (41 sites). The data covered a range of projects in 37 nations from all major landmasses except Antarctica. The variations in costs were compared with a suite of measures of development, and were used to build a simple model for predicting costs elsewhere, and to explore global variation in likely conservation benefits. The findings on conservation costs and benefits were compared with the current global distribution of conservation investment, thus allowing an estimate of global shortfall in funding. The sources of data used by Bruner (2003) have not been established, but probably included existing cost studies, such as used in the 2004 paper (Bruner et al., 2004).

7

All the above studies are limited, being based on different samples of countries due to lack of adequate data; they involve extrapolation to obtain global figures and contain no estimate of margins of error. They all have inherent assumptions, which vary according to the methodology used. Thus a strict comparison is not possible, although the different studies do give estimates of similar orders of magnitude.

Additional costs for an adequate protected area network for climate change adaptation

The greatest limitation for application to costing climate change adaptation in natural ecosystems is that the above calculations are focused on the PAN, which forms only part of the range of adaptation options and does not explicitly include climate change. An additional component therefore is needed. The UNFCCC suggested that this could most easily be done by adding a percentage to current costs of obtaining an improved PA network, but such a figure would be entirely arbitrary, as there are no guidelines as to how this might be estimated and it is likely to vary by country and region according to the sensitivity of the natural ecosystems and the current state of conservation (Berry, 2007). James et al. (2001) did, however, explore two scenarios of expanding the reserve network in regions to 10%, in relation to different levels of IUCN protection categories and hence cost. Once again this involves use of the informed, but arbitrary, 10% figure for scenario development.

The above calculations led to figures that represent the absolute minimum cost of adaptation to climate change, based on improving protection, expanding the network in line with IUCN guidelines, and meeting the opportunity costs of local communities (Table 7.1). An expansion to 10% while maintaining the current ratio of IUCN categories was estimated to involve an annual increase in expenditure of $12.0 billion (scenario 1), while increasing categories I to III to 10% in each region led to increased annual expenditure of $21.5 billion (scenario 2). These calculations are only for the terrestrial PAN and do not include what is currently spent on reserves. Using a similar methodology to James et al. (2001), Balmford et al. (2002) calculated that to cover 30% of the total area of the seas with marine PA would cost at most about $23 billion a year, with about $6 billion a year in starting costs for 30 years.

Costing adaptation outside the protected area network

Costing adaptation in the wider landscape was not part of the overall UNFCCC report and is included only in the consultant's report (Berry, 2007). Again this is based on James et al. (2001), and biodiversity conservation in the wider matrix of landscapes is estimated to cost about $290 billion a year, in addition to the PA network costs (Table 7.2). Such a figure would include more of the adaptation actions identified, although options such as *ex situ* conservation would not be included. Also this estimate is based on extrapolation of figures for the protection of biodiversity on agricultural land from only one study to the global level and thus must be treated with great caution.

7

Table 7.1 Additional adaptation costs under different scenarios

Action	Cost (US$ billion) Mitigation scenario	Cost (US$ billion) BAU scenario
Current global spend on conservation	6	6
Additional spend for expanding terrestrial PAN	12	21.5
Additional spend for expanding terrestrial and marine PAN	41	50.5
Additional spend for expanding terrestrial and marine PAN and surrounding landscape	331	341.5

Table 7.2 Global conservation-related needs for different sectors

Sector	Cost (US$ billion, 1996)
Protecting biodiversity in UK agriculture, extrapolated for global agricultural remediation	240
Forests[1]	34
Marine and coastal areas[1]	14
Freshwater ecosystems[1]	1
Total	**290**

1 From United Nations (1993)

7.4 Case studies

The study closest to identifying actual adaptation costs is for the Netherlands, where it has been estimated that €1 billion are spent on nature conservation, with €285 million for managing national parks and reserves and €280 million for new reserve networks and habitat improvement. This action was aimed at reducing the threat from habitat fragmentation and other sources. The planned national reserve network will reduce the vulnerability of ecosystems and species to climate change and thus a (significant) proportion of the above costs could be considered as climate change adaptation costs.

There are costed case studies of various conservation actions that would contribute to climate change adaptation at national to local scales. In all cases there is the question of what extra action is needed for climate change adaptation. For example, corridors for increasing habitat connectivity are a conservation action also appropriate for adaptation. Naidoo and Ricketts (2006) undertook cost-benefit analysis for the hypothetical construction of three alternative corridors for linking two forest reserves in the Mbaracyau Forest Biosphere Reserve, Paraguay. They showed that the costs varied from $115,000 to $37,000, and although local benefits-costs were negative ($90,000–9000), overall benefits-costs (based on net present value) were positive, ranging from $1.67 million to $1.45 million.

A national example is the costing of the impacts and adaptation for different sectors, including biodiversity, in the UK (Metroeconomica, 2006). This study estimated the annual habitat costs for restoration and re-creation under climate change for 11 important habitats as up to £2.5 million under a 2050s high-emissions scenario. The report pointed out that these figures are 'indicative at best, their lack of robustness reflecting the current lack of monetary data that can generate WTP measures' (Metroeconomica, 2006, pages 6–23).

7.5 Literature

There are no previous explicit assessments of global costs of adaptation in natural ecosystems, but there have been other costings of the various components used and mentioned in the UNFCCC study. Balmford et al. (2002), using similar calculations to James et al. (2001), calculated the annual outlay of properly managing existing PAs and expanding the network to cover 15% of the area of each region as between $20 billion and $28 billion. The Durban Accord from the Fifth World Parks Congress suggested that existing PAs suffer an annual funding gap of some $25 billion, *excluding* additional resources required to expand PA systems. The figures used in the UNFCCC study, therefore, could be considered conservative. Other studies on annual total management costs for PAs in developing countries are $1.1 billion (Vreugdenhil, 2003) and approximately $1.8 billion (Bruner et al., 2004). These are less than the $2.3 billion of James et al. (1999), due to the inclusion of different component costs.

An alternative and complementary approach to costing adaptation may be provided by The Economics of Ecosystems and Biodiversity (TEEB) project.[2] Phase II is undertaking a global study of the economic value of biodiversity loss. If the climate change component of this can be identified, then the costs of avoiding this through adequate and successful adaptation could be calculated.

7

7.6 Discussion and conclusions

The costing of adaptation in natural ecosystems is inherently problematic. While there are guiding principles for adaptation there are few case studies of measures that have been implemented purely as an adaptation to climate change and even fewer have been fully costed. These studies are local to regional and context-specific, making extrapolation to the global scale unwise. The UNFCCC study is the only global one with specific details for natural ecosystems and it is predicated on the assumption that a first step in adaptation is bringing current PAN up to an arbitrary level. It could be argued that this is a cost of conservation rather than adaptation, but given the current state of PAN such expenditure could be viewed as making good the adaptation deficit caused by various drivers.

If the costs of enhancing the conservation of ecosystems in the wider landscape are included, then the UNFCCC figures are a considerable under-estimate of the costs of adaptation, although the basis for these calculations, especially for agriculture, is much more speculative. Care should be taken here that double counting does not occur as some sectoral adaptations can also constitute an adaptation for natural ecosystems. Conversely, others may lead to additional costs.

While all the figures are based on best available data, important assumptions underpin them and considerable gaps exist, leading to the need for extrapolation, sometimes from a limited base. If global costs of adaptation are to be calculated, there is a need to establish what actions truly constitute adaptation to climate change, to identify the limits of adaptation and thus the residual damages, to find ways of valuing the full suite of adaptation measures that can be applied globally and to integrate these costs across sectors.

1 'Natural ecosystems' are defined as systems in which there has been no modification by humans (Calow, 1998).
2 http://ec.europa.eu/environment/nature/call_evidence.htm.

References

Balmford A, Gaston, K.J., Blyth, S., James, A. and Kapos V. |(2003) 'Global variation in terrestrial conservation costs, conservation benefits, and unmet conservation needs', *Proceedings of the National Academy of Sciences USA* 100, pp1046–1050.

Balmford, A., Bruner, A., Cooper, P., Costanza, R., Farber, S., Green, R.E., Jenkins, M., Jefferiss, P., Jessamy, V., Madden, J., Munro, K., Myers, N., Naeem, S., Paavola, J., Rayment, M., Rosendo, S., Roughgarden, J., Trumper, K. and Turner, R.K. (2002) 'Economic reasons for conserving wild nature', *Science* 297, pp950–953.

Berry, P., Paterson, J., Cabeza, M., Dubuis, A., Guisan, A., Jäättelä, L., Kühn, I., Midgley, G., Musche, M., Piper, J. and Wilson, E. (2008) *Mitigation Measures and Adaptation Measures and their Impacts on Biodiversity.* Report to the European Commission as part of the MACIS project (Available from:http://www.macis-project.net/MACIS-deliverable-2.2-2.3.pdf).

Berry, P. (2007) *Adaptation Options on Natural Ecosystems.* A Report to the UNFCCC Secretariat Financial and Technical Support Division (available from: http://unfccc.int/files/cooperation_and_support/financial_mechanism/application/pdf/berry.pdf).

Brooker, R. and Young, J. (2006) *Vulnerability and Adaptation to Climate Change in Europe.* EEA, Copenhagen.

Bruner, A.G. (2003) *How Much will Effective Protected Area Systems Cost?* Conservation International.

Bruner, A., Gullison, R. and Balmford, A. (2004) 'Financial costs and shortfalls of managing and expanding Protected Area systems in developing countries', *Bioscience* 54(12), pp1119–1126.

Calow, P. (ed.) (1998) *The Encyclopaedia of Ecology and Environmental Management.* Blackwell Science, Oxford.

Convention on Biological Diversity (accessed on 30 January 2009) *Integrating Biodiversity into Climate Change Adaptation Planning* (available from http://adaptation.cbd.int/).

Emerton, L., Bishop, J. and Thomas, L. (2006) *Sustainable Financing of Protected Areas: A Global Review of challenges and Options.* IUCN, Gland and Cambridge (x + 97pp).

Fischlin, A., Midgley, G.F., Price, J.T., Leemans, R., Gopal, B., Turley, C., Rounsevell, M.D.A., Dube, O.P., Tarazona, J. and Velichko, A.A. (2007) 'Ecosystems, their properties, goods, and services', in M.L. Parry, O.F. Canziani, J.P. Palutikof, P.J. van der Linden and C.E. Hanson (eds) *Climate Change 2007: Impacts, Adaptation and Vulnerability.* Contribution of Working Group II to the Fourth Assessment Report of the Intergovernmental Panel on Climate Change. Cambridge University Press, Cambridge, pp211–272.

Hopkins, J.J., Allison, H.M., Walmsley, C.A., Gaywood, M. and Thurgate, G. (2007) *Conserving Biodiversity in a Changing Climate: Guidance on Building Capacity to Adapt.* Department for the Environment, Food and Rural Affairs, London.

IPCC (2007) *Climate Change 2007: Impacts, Adaptation and Vulnerability. Contribution of Working Group II to the Fourth Assessment Report of the Intergovernmental Panel on Climate Change*, M.L. Parry, O.F. Canziani, J.P. Palutikof, P.J. van der Linden and C.E. Hanson (eds). Cambridge University Press, Cambridge, 976pp.

IUCN (1993) *Parks for Life: Report of the IVth World Congress on National Parks and Protected Areas.* IUCN, Gland.

James, A., Gaston, K. and Balmford, A. (2001) 'Can we afford to conserve biodiversity?', *BioScience* 51, pp43–52.

James, A.N., Green, M.J.B. and Paine, J.R. (1999). *Global Review of Protected Area Budgets and Staff.* WCMC, Cambridge.

Lapham, N. and Livermore, R. (2003) *Striking a Balance: Ensuring Conservation's Place on the International Biodiversity Assistance Agenda.* Conservation International Center for Applied Biodiversity Science & Center for Conservation and Government, Washington, DC.

Metroeconomica (2006) *Climate Change Impacts and Adaptation: Quantify the Cost of Impacts and Adaptation.* Report to Defra, London (available from http://www.ukcip.org.uk).

Naidoo, R., and Ricketts, T.H. (2006) 'Mapping the economic costs and benefits of conservation', *PLoS Biology* 4(11): e360. DOI: 10.1371/journal.pbio.0040360.

Nordhaus, W.D. and Boyer, J.G. (2000) *Warming the World: the Economics of the Greenhouse Effect.* MIT Press, Cambridge, MA.

Oxfam (2007) *Adapting to Climate Change. What is Needed in Poor Countries and Who Should Pay?* Oxfam Briefing Paper 104, Oxfam, Oxford.

Paterson, J.S., Araújo, M.B., Berry, P.M., Piper, J.M. and Rounsevell, M.D.A.R. (2008) 'Mitigation, adaptation and the threat to biodiversity', *Conservation Biology* 22, pp1352–1355.

Pearce, D. and Palmer, C. (2001) Public and private spending for environmental protection: A cross-country policy analysis', *Fiscal Studies* 22(4), pp403–456.

Pöyry, J. and Toivonen, H. (2005) *Climate Change Adaptation and Biological Diversity.* FINADAPT Working Paper 3, Finnish Environment Institute Mimeographs 333, Helsinki, 46pp.

Stern, N. (2006) *Stern Review: Economics of Climate Change.* Cambridge University Press, Cambridge.

Tol, R.S.J. (2002) 'Estimates of the damage costs of climate change. Part 1: Benchmark estimates', *Environmental and Resource Economics* 21, pp47–73.

UNDP (2007) *Human Development Report 2007/08.* Palgrave Macmillan, New York.

UNFCCC (2007) *Investment and Financial Flows to Address Climate Change.* Climate Change Secretariat, Bonn.

United Nations (1993) *Agenda 21: Rio Declaration and Forest Principles.* Post-Rio edition. New York: United Nations Publications.

Vreugdenhil, D. (2003) 'Modeling the financial needs of protected area systems: An application of the "Minimum Conservation System" design tool', *Fifth World Parks Congress*, 8–17 September, Durban.

World Bank (2006) *Investment Framework for Clean Energy and Development.* World Bank, Washington DC.

8

The costs and benefits of adaptation

Chris Hope

8

Summary

- How worthwhile will adaptation be? In this chapter, we run the PAGE2002 model to calculate the aggregated global costs and benefits of the adaptation measures assumed in the Stern Review – adaptation that costs about the same as assumed in the UNFCC report of 2008.

- The adaptation measures are very worthwhile, with a mean benefit:cost ratio of about 60 in the business-as-usual A2 scenario, and about 20 in the aggressive abatement '450ppm' scenario (to maintain atmospheric levels of carbon dioxide below 450ppm).

- The adaptation measures reduce the mean net present value (NPV) of impacts over the next two centuries by about 28% in the A2 scenario; 72% of the impacts remain even after adaptation. They are a bit more effective in the '450ppm' scenario, reducing the mean impacts by about 33%.

- A more targeted approach to adaptation, with slightly lower amounts in 2020, and greater amounts in 2040 and later, could bring an even higher mean NPV of net benefit than the $300 trillion found in the A2 scenario with the measures assumed in the Stern Review.

8.1 Introduction

The UNFCCC study did not estimate the total value of impacts avoided by adaptation to climate change, so it could not determine whether the benefits of avoided damage exceed the adaptation costs (UNFCCC, 2008, executive summary, paragraph 33). This short chapter begins to address this gap by showing how the costs and benefits of adaptation are treated in a global probabilistic integrated assessment model. It takes as a case study the costs and benefits found by the model used to calculate impacts in the Stern Review (Stern, 2006), the PAGE2002 model, with the same inputs as used in the Stern Review (including a 0.1% pure time preference rate, and an equity weight of 1). The costs and benefits of adaptation are shown for the non-intervention emissions from the A2 scenario. We also investigate the costs and benefits of adaptation in a strict abatement scenario designed to keep atmospheric concentrations of carbon dioxide (CO_2) below 450ppm.

PAGE2002 is a simulation model, estimating the climate consequences and impacts that result from a user-specified emissions scenario. It uses a number of simplified formulas to represent the complex scientific and economic interactions of climate change. A full description of the model can be found in Hope (2006). Most of the model's coefficients and data ranges are calibrated to match the projections of the Third Assessment Report of the Intergovernmental Panel on Climate Change (IPCC, 2001a; IPCC, 2001b). Because many aspects of climate change are subject to uncertainty, PAGE uses probability distributions, based on the best available estimates found in the literature, to represent key inputs to the calculations.

Using these input distributions, PAGE performs a version of Monte Carlo analysis called Latin Hypercube Sampling. The model includes ten time intervals spanning 200 years, divides the world into eight regions, and explicitly considers three different greenhouse gases (carbon dioxide, methane, and sulphur hexafluoride), with other gases included as an excess forcing projection.

8

Three types of impact are calculated:

1 economic impacts, which are impacts on marketed output and income, in sectors such as agriculture and energy use, that are directly included in GDP

2 non-economic impacts, which are impacts on things like health and wilderness areas which are not directly included in GDP

3 discontinuity impacts, which are the increased risks of climate catastrophes, such as the melting of the Greenland or West Antarctic Ice Sheet.

These three types of impacts are summed to calculate total impacts.

The PAGE2002 defaults, adopted by the Stern Review (Stern, 2006), assume that substantial adaptation will occur; as we see below, the annual cost of this adaptation is very similar to the UNFCCC estimates. The reported impacts are for damages remaining after that adaptation takes place. Specifically, PAGE assumes that in developing countries, 50% of economic damages are eliminated by low-cost adaptation. In OECD countries, the assumption is even stronger: 100% of the economic damages resulting from the first 2 degrees of warming, and 90% of economic damages above 2 degrees, are eliminated. For non-economic impacts, adaptation is assumed to remove 25% of the impact everywhere. No adaptation is assumed for discontinuity impacts. Of course, this adaptation policy is simply an assumption, and any other pattern of adaptation can be similarly investigated.

Results are shown as net present values (NPVs) in Tables 8.1 and 8.2, and as annual values in 2060 and other years in Figures 8.1 to 8.9. All results are shown as ranges to enable the uncertainties to be appreciated: 5%, mean and 95% points on the probability distribution in the tables, and the full probability distributions in the figures.

8.2 A2 scenario

Net present value results

Standard economic theory tells us that the NPV results are the main summary of how worthwhile an action such as adaptation is. In the PAGE2002 model, the NPVs of the adaptation policy are calculated by summing the impacts and costs of the adaptation from 2000 to 2200, and discounting them back to 2000. With the Stern Review inputs, the discount rate varies with time and region, but is generally in the range of 1–2% per year.

8

Table 8.1 Net present value of climate change impacts and adaptation costs, A2 scenario

	Trillion US$ (2000)		
	5%	mean	95%
Impacts (with adaptation)	170	890	2340
Impacts (no adaptation)	270	1240	3290
Adaptation costs	4	6	9

Source: 10000 PAGE2002 model runs with 'Stern Review' assumptions

Table 8.1 shows that adaptation reduces the mean net present value of impacts in the A2 scenario by $350 trillion, about 30%, from $1240 trillion to $890 trillion. About 72% of the impacts remain even after adaptation, that costs a very similar amount to the costs estimated by UNFCCC (2008). The mean cost of the adaptation measures is $6 trillion. Even though the adaptation does not remove all, or even most, of the impacts it is clearly very worthwhile, with a mean net benefit of well over $300 trillion, at a mean benefit:cost ratio of about 60. This is the main result of the modelling, but looking in more detail at the annual costs and benefits can help to explain how this result occurs.

Annual costs and benefits

Results are shown here primarily for 2060, the analysis year closest to the middle of the century. Results for 2020, 2040, 2080 and 2100 are also briefly summarised below. Figure 8.1 shows that the mean impacts in 2060 are about $1.5 trillion (the horizontal axis of the figure says 'millions', but the currency unit in the model is millions of dollars, so this is millions of millions, or trillions), with a 5–95% range of about $0.3–3.6 trillion. As usual, there is a long right tail, with a small probability of impacts as large as $20 trillion.

Figure 8.1 Impacts in 2060, A2 scenario

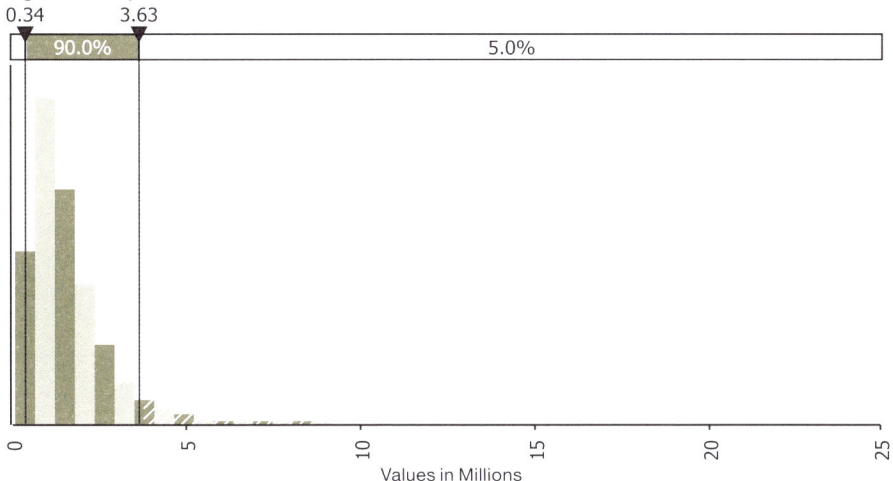

Source: 10000 PAGE2002 model runs with 'Stern Review' assumptions

8

Figure 8.2 shows that the mean impacts in 2060 without adaptation are about $2.4 trillion, with a 5–95% range of about $0.6–5.6 trillion. The long right tail extends out to over $25 trillion.

Figure 8.2 Impacts with no adaptation in 2060, A2 scenario

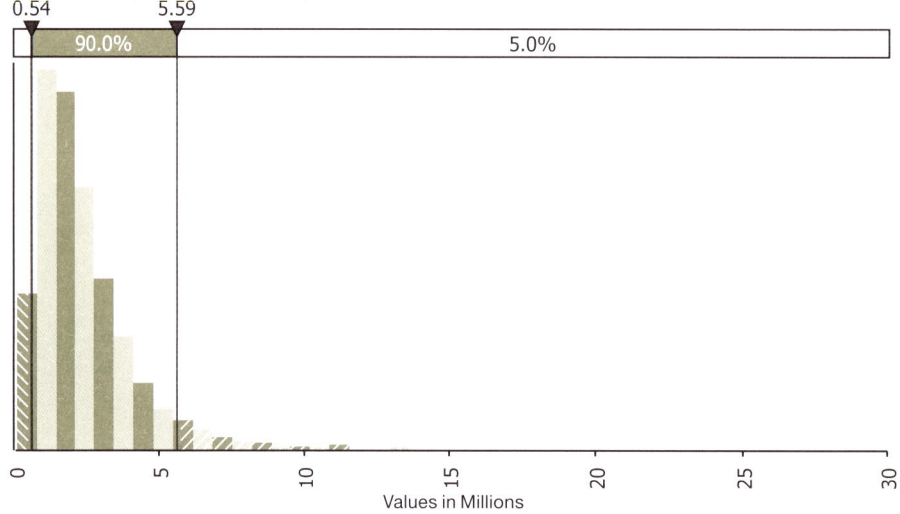

Source: 10000 PAGE2002 model runs with 'Stern Review' assumptions

Figure 8.3 shows the resulting benefit of adaptation in 2060, which has a mean value of about $0.8 trillion ($2.4 trillion minus $1.5 trillion ignoring rounding errors) and a 5 – 95% range of about $0.2 – 1.9 trillion.

Figure 8.3 Benefit of adaptation in 2060, A2 scenario

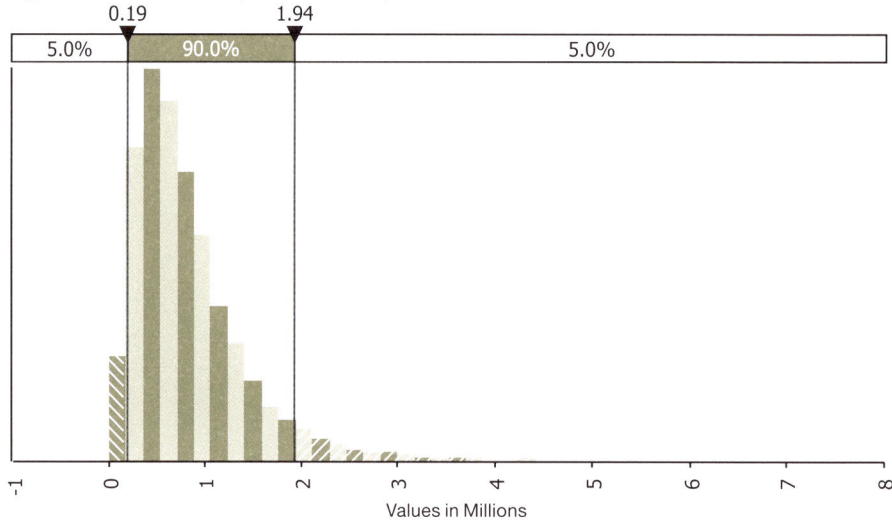

Source: 10000 PAGE2002 model runs with 'Stern Review' assumptions

Figure 8.4 shows the main influences on the benefit of adaptation in 2060. The values plotted in Figure 8.4 are the amount by which the benefit of adaptation in 2060 would increase if the input in question increased its value by one standard deviation. The eight impacts shown, in order of importance, are:

1 Climate sensitivity, which is assumed to have a range of 1.5–5°C, with a most likely value of 2.5°C. Increasing this by one standard deviation would worsen the climate change problem and increase the benefit of adaptation in 2060 by about $300 billion.
2 The economic impact in the focus region, the EU, which is assumed to have a range of minus 0.1% (a slight benefit) to 1% of GDP for a 2.5°C temperature rise, with a most likely value of 0.6% of GDP.
3 The indirect sulfate effect which counteracts the effect of greenhouse gases to some extent. The higher this parameter, the smaller the indirect sulfate effect, so the more serious climate change will be, and the larger the benefit from adaptation.
4 The half-life of global warming, or the time the mean global temperature takes to respond to a change in greenhouse gas concentrations. The larger this parameter, the further in the future the main climate change impacts occur, and so the smaller the benefit of adaptation in 2060.
5 The non-economic impact parameter. The adaptation policy is assumed to be more effective against economic than non-economic impacts, so it is no surprise that the non-economic impact parameter is less influential.
6 The relative weight applied to impacts in the India region; this region is the one that is assumed to be hardest hit by climate change in the default set-up of the model.
7 Another scientific parameter describing how much of the CO_2 emissions escape the very short-term removal processes of the biosphere and oceans.
8 The weight applied to impacts in another vulnerable region, Africa.

Four of these top eight influences are scientific and four are economic, which shows the importance of taking an integrated approach to the analysis.

Figure 8.4 Main influences on the benefit of adaptation in 2060, A2 scenario

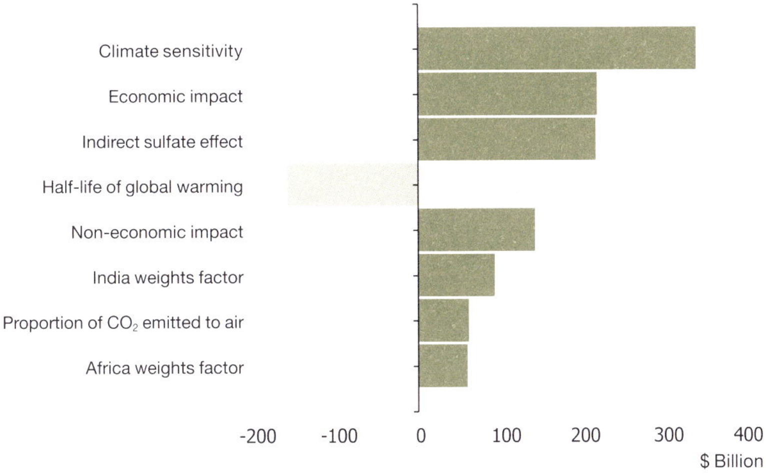

Source: 10000 PAGE2002 model runs with 'Stern Review' assumptions

8

Figure 8.5 shows that the mean cost of adaptation in 2060 is about $90 billion ($0.09 trillion) (the horizontal axis of the figure says 'thousands', but the currency unit in the model is millions of dollars, so this is thousands of million, or billions), with a 5–95% range of about $58–132 billion, similar to the UNFCCC estimates. There is a range not because the amount of adaptation varies, but because the unit cost of adaptation measures is uncertain. Thus in 2060 in the A2 scenario, adaptation at a mean cost of about $0.09 trillion brings a mean benefit of about $0.8 trillion, giving a mean net benefit of about $0.7 trillion at a mean benefit:cost ratio of nearly 10:1.

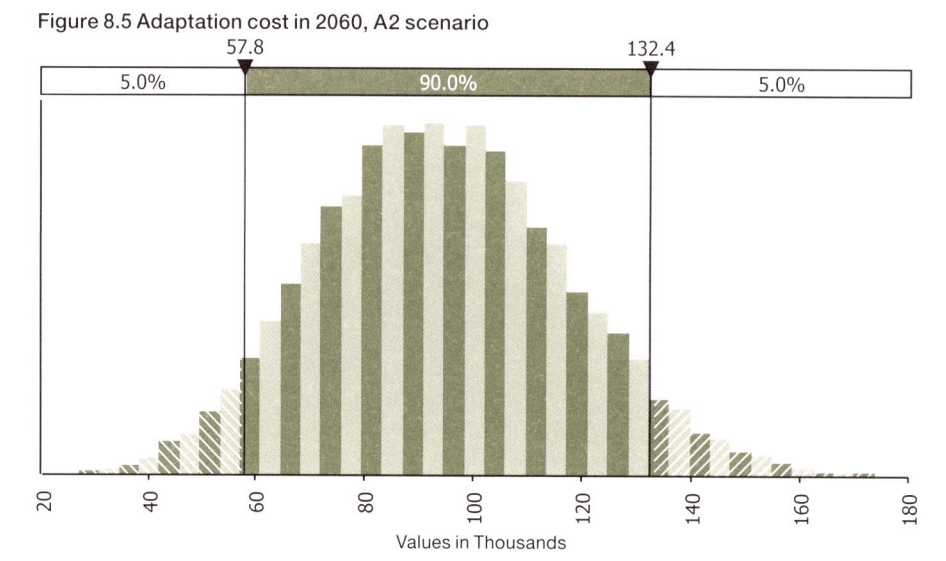

Figure 8.5 Adaptation cost in 2060, A2 scenario

Source: 10000 PAGE2002 model runs with 'Stern Review' assumptions

Costs and benefits from 2020 to 2100

Under the Stern Review assumptions, the annual cost of adaptation is the same in each year after 2020, when the full set of adaptation measures are assumed to be implemented. Figure 8.6 shows the mean benefit of adaptation from 2020 to 2100 (thicker line), and the 5 and 95% points (thinner lines). The mean benefit is about $60 billion in 2020, $260 billion in 2040, $2.2 trillion in 2080 and $5.4 trillion in 2100. In 2020 the mean benefit is less than the mean cost of about $90 billion, giving a mean net benefit of about –$30 billion, and a mean benefit:cost ratio of about 0.7. On the other hand, mean benefit:cost ratios in 2080 and 2100 are even higher than in 2060. These results suggest that a more targeted adaptation policy, with slightly lower amounts in 2020, and greater amounts in 2040 and later, could bring an even higher mean NPV of net benefit than the $300 trillion or so found with the measures assumed in the Stern Review.

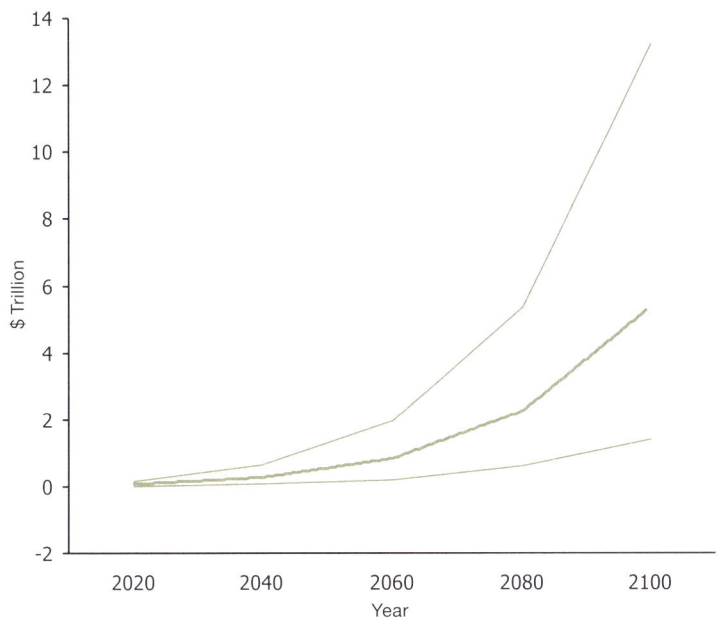

Figure 8.6 Benefit of adaptation from 2020 to 2100, A2 scenario

Source: 10000 PAGE2002 model runs with 'Stern Review' assumptions

8.3 The '450ppm' scenario

If adaptation is very worthwhile under business-as-usual emissions, as the previous results have shown, what if an aggressive abatement policy is followed? Is adaptation still as desirable? This can be investigated by re-running the adaptation policy assessment with a more aggressive abatement policy in the PAGE2002 model.

Net present value results

Table 8.2 Net present value of climate change impacts and adaptation costs, '450ppm' scenario

	Trillion US$ (2000)		
	5%	mean	95%
Impacts (with adaptation)	60	275	760
Impacts (no adaptation)	100	410	1070
Adaptation costs	4	6	9
Abatement costs	50	110	170

Source: 10000 PAGE2002 model runs with 'Stern Review' assumptions

8

Table 8.2 shows that the impacts are much lower under the '450ppm' scenario, as would be expected. (The scenario is described as '450ppm' (in quotes), rather than 450ppm, because the carbon cycle feedbacks in the PAGE2002 model mean that the CO_2 concentrations actually do not stay below 450ppm; their mean value reaches about 500ppm by 2100, but this is still well below the mean value of over 800ppm reached by 2100 in the A2 scenario.) Adaptation reduces the mean net present value of the impacts by $135 trillion, from $ 410 to 275 trillion. About 67% of the impacts remain even after adaptation, a slightly lower proportion than under the A2 scenario, as adaptation is more effective against the smaller climate changes in this aggressive abatement scenario.

The mean cost of the adaptation measures is $6 trillion. The adaptation is clearly still very worthwhile even if aggressive abatement measures are introduced, with a mean net benefit of about $130 trillion, at a mean benefit:cost ratio of over 20. Again, it is worth a brief glimpse at the annual costs and benefits, as follows, to understand better this main result.

Annual costs and benefits

Figure 8.7 shows that the mean impacts in 2060 are about $1.2 trillion, with a 5–95% range of about $0.3–2.8 trillion. The long right tail extends out to over $10 trillion. These impacts in 2060 are only about 20% below the impacts in the A2 scenario; the main differences in climate outcomes between the scenarios come in the 22nd century.

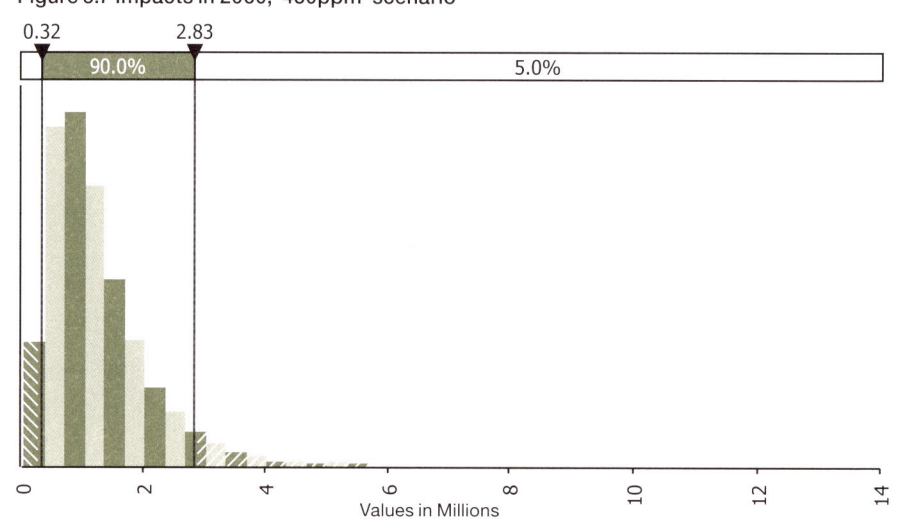

Figure 8.7 Impacts in 2060, '450ppm' scenario

Source: 10000 PAGE2002 model runs with 'Stern Review' assumptions

Figure 8.8 shows that the mean impacts in 2060 without adaptation are about $1.9 trillion, with a 5–95% range of about $0.5–4.4 trillion. The long right tail extends out to over $15 trillion.

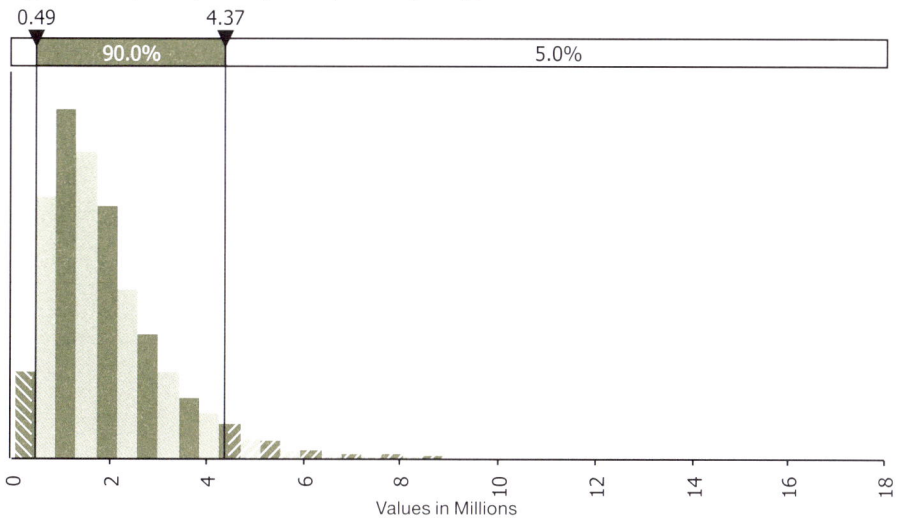

Figure 8.8 Impacts (no adaptation) in 2060, '450ppm' scenario

Values in Millions

Source: 10000 PAGE2002 model runs with 'Stern Review' assumptions

Figure 8.9 shows the mean cost of adaptation in 2060 is about $90 billion ($0.09 trillion), with a 5–95% range of about $55 to $130 billion. This is identical to the A2 scenario, as identical adaptation measures are assumed to be in place.

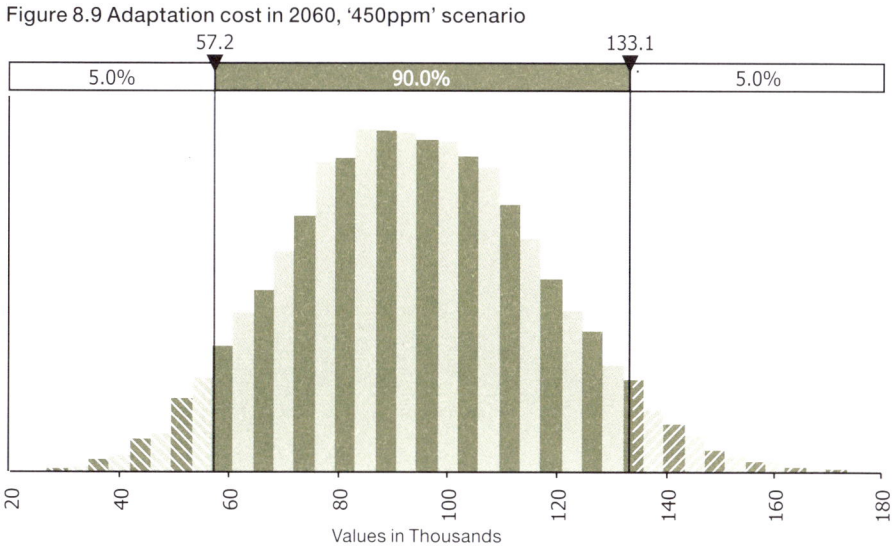

Figure 8.9 Adaptation cost in 2060, '450ppm' scenario

Values in Thousands

Source: 10000 PAGE2002 model runs with 'Stern Review' assumptions

Assessing the costs of adaptation to climate change | 109

8

Thus in 2060 in the '450ppm' scenario, adaptation at a mean cost of about $0.09 trillion brings a mean benefit of about $0.7 trillion, giving a mean net benefit of about $0.6 trillion at a mean benefit:cost ratio of nearly 8:1. As in the A2 scenario, this benefit:cost ratio will increase rapidly in later years.

8

References

Hope, C. (2006). 'The Marginal Impact of CO2 from PAGE2002: An Integrated Assessment Model Incorporating the IPCC's Five Reasons for Concern.'

IPCC (2001a). *The Scientific Basis*. Contribution of Working Group I to the Third Assessment Report of the Intergovernmental Panel on Climate Change, Cambridge University Press, Cambridge.

IPCC (2001b) *Impacts, Adaptation and Vulnerability*. Contribution of Working Group II to the Third Assessment Report of the Intergovernmental Panel on Climate Change, Cambridge University Press, Cambridge.

IPCC (2007). *Climate Change 2007: The Physical Science Basis*. Contribution of Working Group I to the Fourth Assessment Report of the Intergovernmental Panel on Climate Change. Cambridge University Press, Cambridge.

Stern, N. (2006). *The Stern Review: The Economics of Climate Change*. HM Treasury, London.

UNFCCC (2008). *Investment and Financial Flows to Address Climate Change*. UNFCCC, Bonn.